Gordon Burne

Engines

Illustrated by John Plumb

Wayland

Endeavour Books

Initiated and edited by Stella Robinson, the books in the Endeavour series have been researched and written by qualified scientists and engineers and each title covers one aspect of man's scientific and technological development.

With the help of biographical notes, full colour illustrations and explanatory diagrams, each book highlights the major personalities, discoveries and developments in a particular field of scientific or technological progress.

Ranging in scope from earliest times to the present day, the series provides a comprehensive account of man's struggle to understand the natural world and organize the environment to his benefit; man's scientific endeavour.

Other titles in the series

Textiles
Buildings
Farming
Health and Medicine
Tools and Manufacturing
Mathematics and Computers
Iron and Steel
Resources

First published in 1985 by Wayland (Publishers) Limited
49 Landsdowne Place, Brighton, East Sussex BN3 1HF, England

ISBN 0 85078 349 6

Designed and Typeset by DP Press, Sevenoaks, Kent
Printed in Italy by G. Canale & C.S.p.A., Turin

Contents

Introduction 4
Early days 6
Thomas Newcomen – the first steam engine 8
James Watt 10
Richard Trevithick 13
Steam at sea and on the land 14
Locomotives 16
Compound engines 18
Boilers 19
The theory of heat 20
The steam engine in its prime 21
Steam turbines 22
Hot air engines 24
Gas engines 26
Petrol engines 28
Diesels 32
Modern diesel engines 34
Aero engines 35
Gas turbines 36
Rocket motors 38
Unusual engines 40
The future 41
Time chart 42
Glossary 46
Index 48

Cover: an American Corliss steam engine of 1876 which gave nearly 2000 kW of power.

Title page: this Swiss Sulzer marine diesel engine is as big as a three-storey house.

Introduction

A steam engine of 1903. Ropes in the grooves of the flywheel drove hundreds of machines in a textile mill.

Like many other words, the word "engine" has had different meanings at different times. At first, it meant "the ability to invent things"; then it came to mean any ingenious mechanical invention, as in "fire-engine" or "*beer-engine*" (the big catapults used by Roman soldiers were called engines). Nowadays the word is normally used to mean a mechanical device for producing power by means of heat, and this is the meaning used in this book. Other ways of producing power, such as water wheels and windmills, will be described in "Energy" in this series.

The first engines, in this sense of the word, were steam engines, with a *piston* which was moved from one end of a *cylinder* to the other by means of steam. They came into use in the early 18th century, at the beginning of the Industrial Revolution. During this time Britain changed from farming to manufacturing; the first country in the world to do so. Factories needed power, and they needed iron and steel; so more and more coal was required to replace the timber that became increasingly scarce. As coal mines became deeper,

The Supermarine S6b, fitted with a Rolls Royce petrol engine, held the world speed record in 1931, and led to the Spitfire of World War II.

This powerful Canadian National diesel-electric locomotive is in three parts so that it can follow the curves of the track.

it was more difficult to keep them from flooding with water, so the first steam engines were used for pumping water out of mines.

Another type of engine used hot air instead of steam in the cylinder. In both types, the reciprocating (to-and-fro) motion of the piston in the cylinder was often turned into rotating motion by means of a *crankshaft*. In the steam *turbine*, developed towards the end of the 19th century, the piston and cylinder were done away with and steam was used to produce rotating motion directly, in the same sort of way as a water wheel or windmill.

In all these engines, fuel was burned outside the engine in a separate furnace. However, it is simpler and more economical to burn the fuel inside the engine itself. Coal could not be used in this way, but oil, gas and petrol can, and *internal-combustion engines* using these fuels started to come into use towards the end of the 19th century. The fuel can be made to explode in the cylinder by means of an electric spark, a hot tube, or a flame; or, if the fuel is compressed very hard, it will become hot enough to set itself on fire, as in the *diesel engine*. Very powerful petrol engines were developed to drive the aircraft of World War II; they produced so much power that it became difficult to design propellers able to use it all. This problem was solved by the invention of the *gas turbine*, in which the fuel is burned inside the engine – an internal-combustion turbine in fact.

A gas turbine-engined Hawk of the RAF Red Arrows aerobatic team.

Part of the steam turbine in a modern power station. It develops a thousand times the power of the mill engine opposite.

Early days

Although piston engines for practical use were developed long before turbines, turbines are much simpler devices, and the first heat engines were turbines. In the 1st century AD, Hero of Alexandria wrote about mathematics, surveying and light. He invented several ingenious devices, such as automatically opening temple doors and the "aeolipile". This was the first steam turbine, which worked on the same principle as many modern garden sprinklers but used steam instead of water.

Hero's aeolipile. Steam from the boiler went up through the supporting tubes into the rotating sphere.

In about AD 1550, an Arab, Taqi-al-Din of Damascus wrote a book called "The Sublime Methods of Spiritual Machines" in which he described how a spit for roasting meat over a fire could be turned by directing a jet of steam on to a wheel with blades. In 1629 Giovanni Branca suggested the turbine shown; it seems very strange nowadays to make the boiler in the form of a man's head.

Branca's turbine.

In the early days, people did not know how to make boilers that would withstand high pressures without bursting, and the first engines worked by *atmospheric pressure*; to use this, it is necessary to create a *vacuum* for the pressure to work against. In 1654, Otto von Guericke, a German military engineer who became Mayor of Magdeburg, demonstrated how great atmospheric pressure is. Two teams of eight horses failed to separate the two halves of a sphere from which the air had been exhausted, so showing that they were pressed together by the atmosphere. Forces nowadays are measured in newtons, a rather small unit, since a medium-sized apple weighs about a newton; so kilonewtons (1 kN = 1000 newtons) are usually used. An average car weighs about 10 kN.

Thomas Savery was born in Devon in the 1650s; nobody seems sure whether he was called "Captain" because he had been the captain of a ship, or a military engineer or the captain of a mine (the managers of Cornish mines were called captains in those days). At any rate, he wrote a book on navigation, and another on fortifications. He demonstrated his engine to King William III in 1698 and took out a patent for it. In 1702, he set up a workshop off Fleet Street in London to build the engines, and advertised them to "The Gentlemen Adventurers in the Mines of England". He also demonstrated a model of his invention to the Royal Society, which made him a Fellow in 1699. Later, he helped to look after wounded sailors and was in charge of the waterworks at Hampton Court Palace. He died in 1715.

Von Guericke's demonstration of atmospheric pressure.

How Savery's pump worked.

Steam from the boiler drives the water out of A through valve B and up the pipe. Then cold water poured on to A creates a vacuum, sucking water up through C to refill A.

The pressure of the atmosphere varies from time to time and from place to place; an average value is 100 kilonewtons – about the weight of a double decker bus – acting on an area of one square metre (100 kN/m^2). A pressure of 100 kN/m^2 is often referred to by engineers as 1 *bar* (the weatherman calls it 1000 millibars). Atmospheric pressure will support a column of water 10 m high or, in a barometer, a column of mercury about 76 cm high.

When water turns into steam, its volume increases by hundreds of times; when steam in a closed container is cooled so that it turns back into water, a vacuum is formed. Captain Thomas Savery was the first to use steam, a vacuum and the pressure of the atmosphere for a really useful purpose – pumping water. He called his machine "The Miner's Friend for raising of water by the impellant force of fire" or, for short, a "Fire Engine". It wasn't really what would be thought of as an engine nowadays; it had no pistons or *rotors*, and the only moving parts were the *valves*. Pumps working on this principle are still in use; they are called pulsometer pumps. Savery's pumps don't seem to have been used much in mines. They would only lift water about 15 m and many mines were over 100 m deep. Besides, it isn't very safe to have a big fire in a mine and the boilers were apt to burst. However, Savery's pumps were certainly used to supply water to water mills and to country houses.

How Savery's engine was installed. With two vessels working alternately the stream of water produced was almost constant.

Von Guericke could have tied one half of the sphere to a tree, and used one team of eight horses, but it might not have been so impressive.

Thomas Newcomen – the first steam engine

The first real steam engine, using a piston in a cylinder, was invented by Thomas Newcomen of Dartmouth. He may have got the idea from Denis Papin (1647–1712), a Protestant refugee from Paris, who invented a pressure cooker which included a *safety valve* kept closed by means of a lever carrying a weight, a type that continued in use for many years. Papin also showed that if steam is used to raise a piston in a cylinder, and the cylinder is cooled so that the steam *condenses* back into water, leaving a vacuum, atmospheric pressure on the top of the piston can be used to raise a large weight.

Papin was a scientist rather than an engineer, and didn't develop his idea any further. Newcomen's engine, which a newspaper called "that surprising machine for raising water

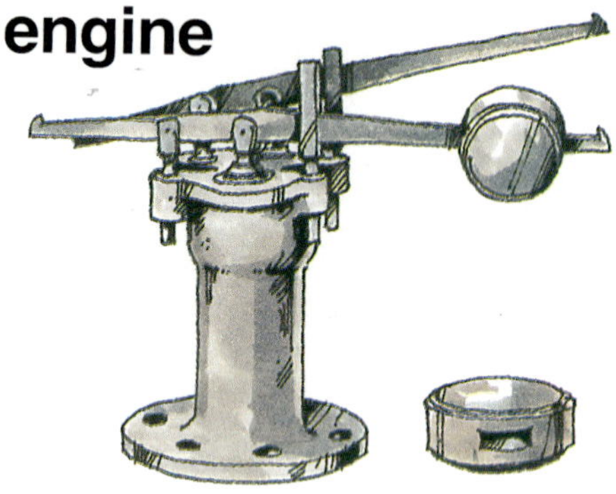

A safety valve of the 19th century. When the steam pressure was too high, it lifted the weighted lever. The position of the weight could be adjusted for different pressures.

A Newcomen engine of 1712.

Thomas Newcomen was born in Dartmouth in 1663. His ancestors were landowners, but the family must have come upon hard times because he started work as a blacksmith making small tools. He sold these to the mines of Devon and Cornwall, thus learning of the need for a powerful pumping engine. Unfortunately, Savery had obtained a patent for "raising water by fire" which prevented Newcomen from selling his engines except in partnership with Savery, though it doesn't seem that Savery had much to do with Newcomen's invention. Newcomen started working on his idea in about 1698; but it wasn't until 1712 that the first one went into action near Dudley in Staffordshire. It replaced a stable of 500 horses previously used to drive the pumps. Newcomen died in London in 1729. Within 40 years of his death, his engines were being used in America, Russia and the Continent as well as in Britain; some continued working into the present century.

by fire", solved the problem of keeping the coal mines free of water. Coal became more freely available and could be used in the form of coke, in blast furnaces, to produce large quantities of iron (see "Iron & Steel" in this series); machines and engines could then be made from iron, much more cheaply than before.

Thus began the Industrial Revolution which transformed Britain from a farming to a manufacturing nation. Newcomen's first engines were made of wood and brass and leather and brickwork; but later huge cast iron cylinders for Newcomen engines were made by Abraham Darby and his family at their famous ironworks in Coalbrookdale.

Savery's machine sucked the water from above, but Newcomen's engines lifted it from beneath so it could raise water through a much greater height. They were so widely used that they soon became known as "common engines". At first, the valves, or cocks, were opened and closed by a "cock boy" and there is a story that one of them found a way of making his work easier by using a piece of string tied to the beam. This seems an unlikely story, because the valve gear which Newcomen later developed had to be very complicated to work properly.

By modern standards, of course, the "fire engine" was very slow – 16 *strokes* a minute at the most – and very inefficient.

John Smeaton (1724–1792) was the first scientific engineer, famous for building the Eddystone Lighthouse, measuring the performance of windmills and water mills, and building canals and harbours. He worked out a way of measuring the efficiency of the Newcomen engine by calculating what he called its duty – the amount of water, in millions of pounds, which could be raised one foot by burning a bushel of coal (34 kg). By making some apparently small improvements, he more than doubled its efficiency; it remained, though, some 30 times less efficient than steam engines of later days.

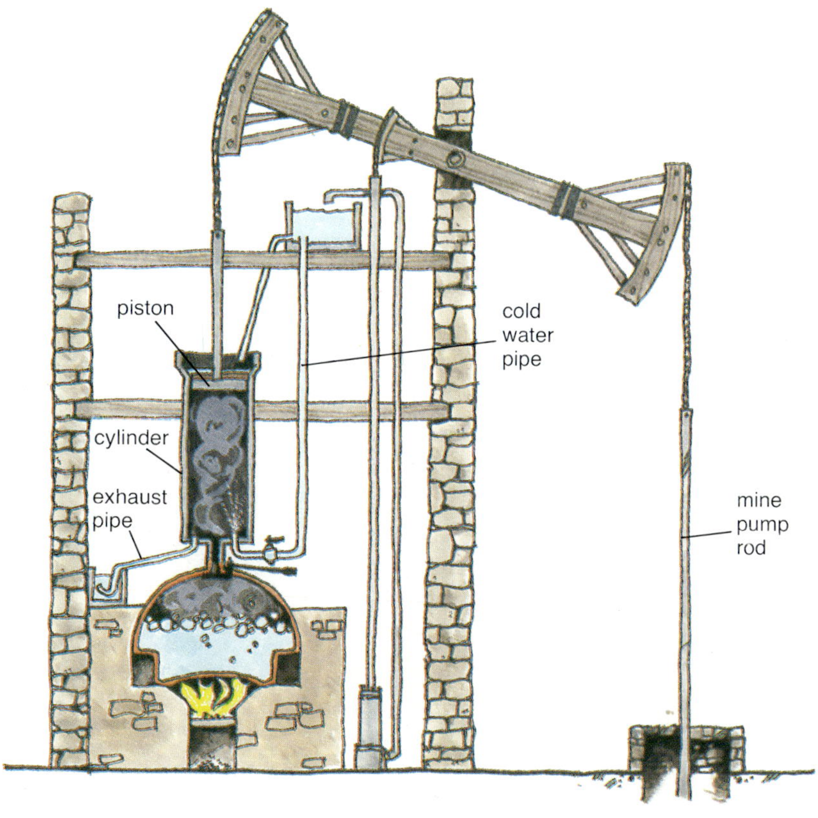

The weight of the mine pump rod on the right pulled the piston up, filling the cylinder with steam. Cold water at the bottom of the cylinder then condensed the steam, and atmospheric pressure pushed the piston down, lifting the pump rod. The exhaust pipe drained the water out of the cylinder.

James Watt

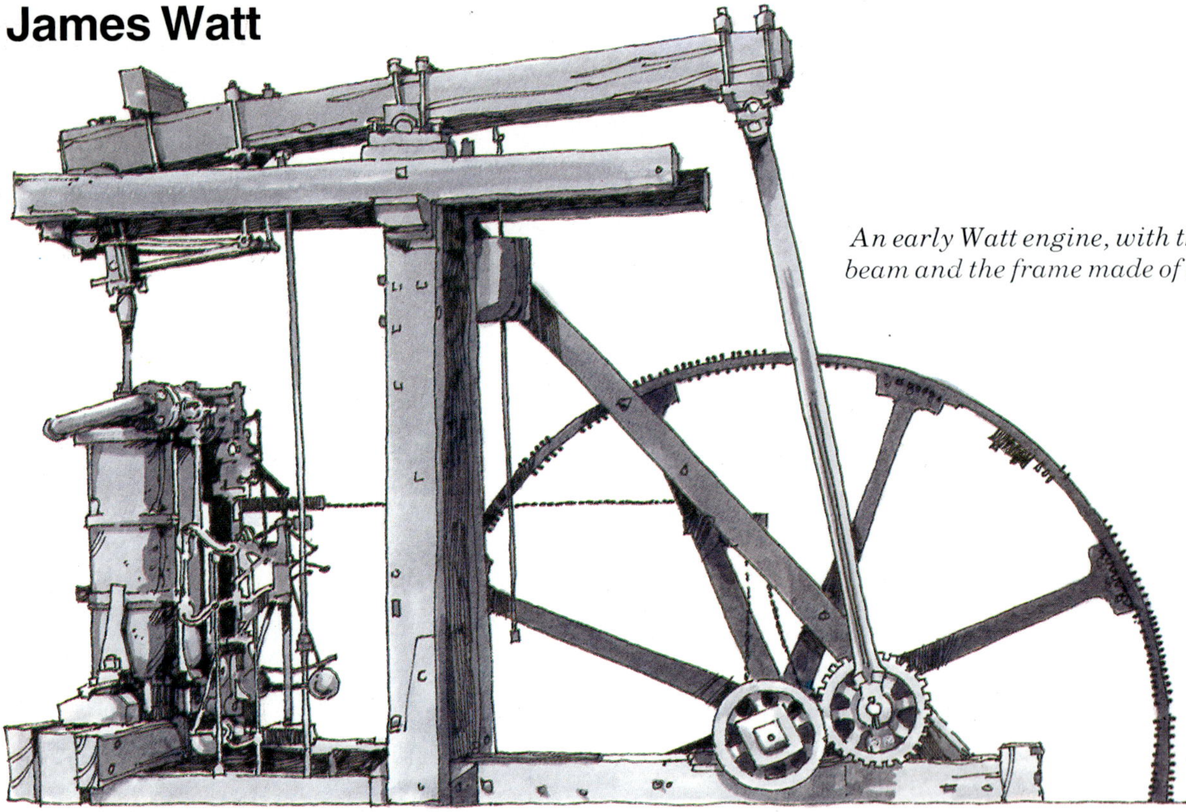

An early Watt engine, with the beam and the frame made of wood.

Newcomen's engines used so much coal that most of them were used in coal mines where coal was relatively cheap. James Watt's great contribution was to make an engine that used only a third as much coal as Smeaton's improved Newcomen engine. Steam engines became so much cheaper to run that they could be used almost anywhere, and for running machinery as well as for pumping water. Watt had to overhaul a model Newcomen engine, and found that it used much more steam and much more cold injection water than he expected. As the steam entered the cold cylinder, much of it condensed into water and much of it was used in heating up the cylinder. Similarly, as the cold water was sprayed into the hot cylinder, much of it evaporated into steam and much of it was used to cool down the cylinder.

Watt experimented for several years to find out how much steam was obtained from a given volume of water. His very simple apparatus made it 1800 times as much; 1600 is a more accurate result. He also found that a great deal of heat was needed to turn water into steam. A kettle which brings water to the boil in 5 minutes would take another 30 minutes to turn all the water into steam. Watt certainly used ordinary kettles in some of his experiments (he also used one of Papin's pressure cookers) but the story that he got his great idea whilst watching a kettle boil is not true; he said himself that the idea came to him when he was going

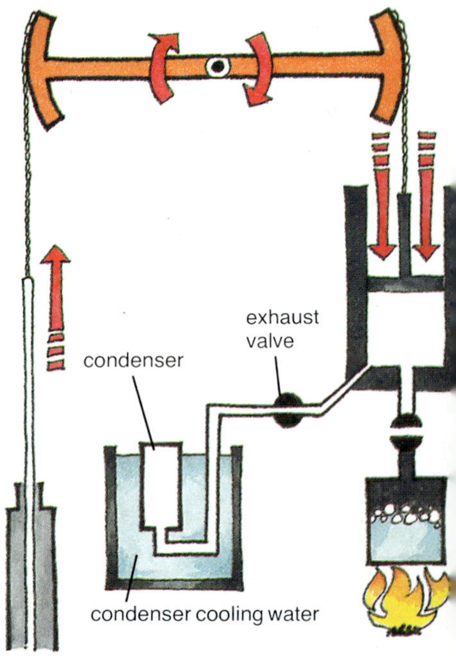

Atmospheric Watt engine. It worked like the Newcomen engine, except that after the piston was raised by the pump rod and the exhaust valve was opened, the steam was condensed in the condenser instead of the cylinder.

James Watt's workshop in Birmingham.

James Watt (1736–1819) trained as a scientific instrument maker, and worked for the University of Glasgow. He was so intelligent that he soon became friendly with the professors and discussed his ideas about heat with them. His first engine, built in 1769 in partnership with John Roebuck of the famous Carron ironworks, was not a success. He worked as a civil engineer on canals for a while until 1774 when he went into partnership with Matthew Boulton of Birmingham and the famous firm of Boulton and Watt was founded. Their first engines were made in 1776 and by 1800, when Watt retired, the firm had made 496 engines. 308 of these were *rotative engines* for driving various types of machines.

for a walk on Glasgow Green, between "the old washing house and the Golf-house". He saw that there was no need to keep on heating and cooling the cylinder. Instead, the cylinder could be kept hot all the time, if the steam was allowed to flow out of the cylinder into a separate vessel which was kept cold all the time, and in which the steam would be condensed into water. This second vessel he called a *condenser*. An air pump was used to create a vacuum in the condenser so that the steam would flow into it quickly.

James Watt made many other improvements, which resulted eventually in much more efficient steam engines. He used a steam *jacket* round the cylinder to keep it hot, and closed off the end of the cylinder, so that the engine no longer relied on atmospheric pressure working against a vacuum, but on a steam pressure greater than that of the atmosphere. The supply of steam could be cut off after the piston had moved a short distance, the expansion of the steam providing the power during the rest of the stroke. Also, steam could be admitted to both ends of the cylinder in turn, making the engine *double acting* and therefore twice as powerful.

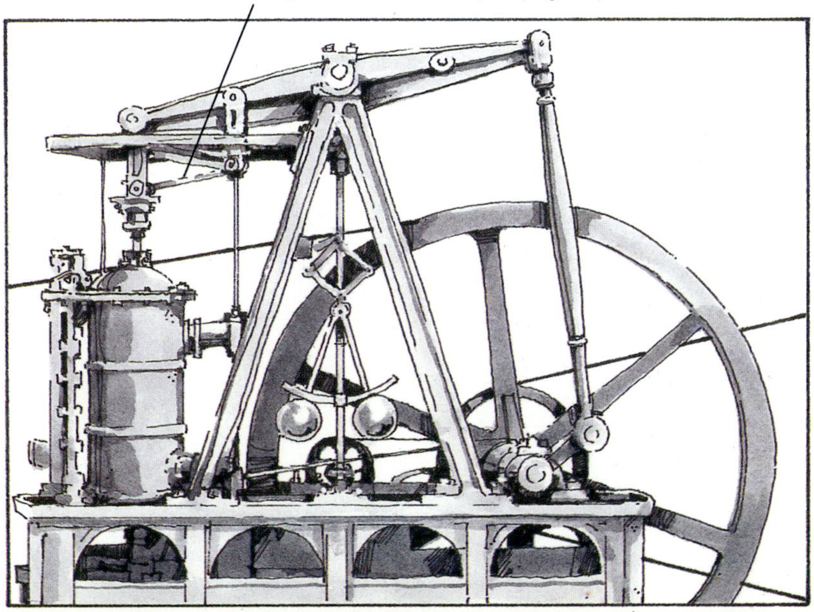

A Watt engine of 1800, the result of thirty years' improvements.

The valve gear of a Watt engine was very complicated.

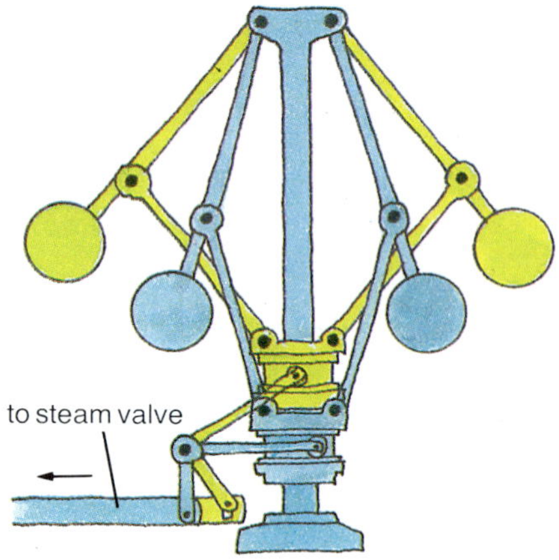

Watt's governor. When the speed of the engine increased, the balls moved outwards and worked a mechanism to close the steam valve.

Now it was no longer possible to hang the piston on a chain from the beam, because it wouldn't work on the upstroke, so Watt invented a mechanism to allow the piston rod to move up and down in a straight line while the end of the beam moved in a circular arc. Watt's engines were more expensive to make than Newcomen's, so the latter continued to be used in the coalfields for many years.

A big problem was to get the huge cylinders made accurately enough for the pistons to fit properly; this problem was solved by "Iron-mad" John Wilkinson of Coalbrookdale (see "Iron & Steel" in this series). Watt advertised his engines by saying how many horses they could replace – he measured how fast a horse could work and then added a bit on so that a 20 *horsepower* Watt engine gave more power than 20 horses. The customer was sure to be satisfied. Nowadays the unit of power is called the *watt*, though this is more often quoted in kilowatts (kW). One horsepower is nearly three quarters of a kilowatt. Watt also developed a *governor* to keep the engine running at constant speed even when the load changed and an engine indicator to measure the variation in steam pressure in the cylinder while the engine was running.

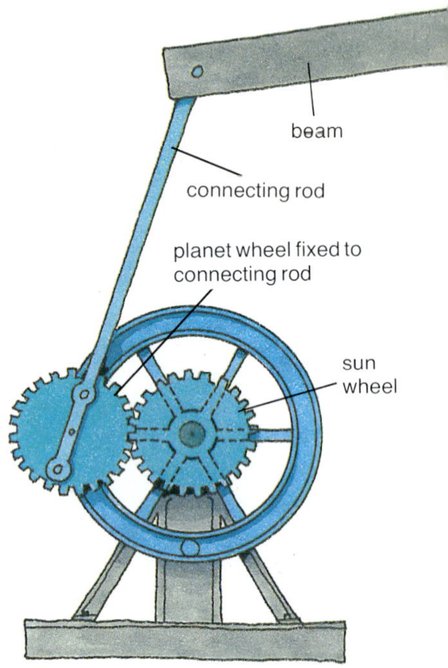

Watt wouldn't use a crankshaft on his early machines because the idea had been patented. Instead, he used the 'sun and planet' gear system to change the up-and-down motion of the beam into rotation of the sun wheel.

Richard Trevithick

Trevithick's horizontal high pressure steam engine of 1803. The cylinder is in the boiler to keep it hot.

Richard Trevithick, was a big, strong, likeable, high-spirited man who seemed to lose all his money as soon as he made it. He was made bankrupt in 1811 and went to South America to work, but became involved in wars, revolutions and crossing the jungles of Central America on foot and by raft, making and losing money. Meanwhile, his wife stayed patiently in Cornwall running a hotel. He returned to England in 1827 with no money. They moved to Dartford where he died in 1833. His fellow workers had to club together to pay for his funeral.

Watt's engines were enormous, heavy and slow; so slow that the engine boy could go along the beam to oil the pivots while the engine was working. They were said to make "unearthly groanings, sighings and yawnings". When Watt's patent ran out in 1800, many people tried to improve on them; the most important of these was Richard Trevithick of Cornwall; known as 'Captain Dick', he was born in 1771, the son of a Cornish mine captain. He made much smaller, lighter engines by using high pressure steam. Watt used steam at about 30 kN/m^2 (0.3 bar); but Trevithick used 3½ bar at first, and later, 10 bar. Watt was against the idea; he said that "strong steam" was dangerous, and he was right; hundreds of boilers exploded during the first half of the 19th century, until the use of steel instead of iron made them much safer. But Trevithick was right as well; high pressure steam was essential for many purposes and resulted in much more efficient engines. Its use meant that the condenser could be done away with (though they are still used in ships and in power stations) and the steam allowed to exhaust to the atmosphere with a characteristic "puffing" sound. Trevithick also thought of letting out this exhaust steam through the chimney. This provided a strong draught for the furnace, making it burn much better. Between 1800 and 1804 he had built 50 steam engines and he made locomotives, a steam dredger and a threshing machine.

Similar improvements to the Watt engine were made in America by Oliver Evans (1755–1819), although at the time there were only half a dozen Watt engines in the whole country.

Another type of Trevithick engine, with a vertical cylinder in the boiler.

Steam at sea and on the land

Over the years, many improvements were made to Watt's type of pumping engine, particularly by the use of higher steam pressures pioneered by Trevithick. So many were made in Cornwall and used there in the lead, tin and copper mines that they came to be known as *Cornish engines* – even when similar engines were made or built elsewhere. They were still being built in the present century, and several of these majestic engines can be seen in action in various museums.

The development of the rotating engine meant that many other uses apart from pumping became possible, in factories, in farming and in transport. The *beam engine* couldn't be used for land transport; it was too big and heavy; but it could be used in ships. Even so, quite a small engine needed quite a large boat to carry it. Papin and Savery had both suggested using engines in ships at the beginning of the 18th century, but it was another 75 years before the idea was tried out in France. The first steam boat that could actually do anything useful was "Charlotte Dundas II", used as a tug on the Forth & Clyde canal in 1803 (see "Iron & Steel" in this series). The engine was developed by William Symington. The trials of the Charlotte Dundas were watched by an American artist and engineer, Robert Fulton, whose "Clermont" of 1806 was the first commercially successful steamer; it sailed the River Hudson from New York to Albany. Fulton had spent 14 years in England and 9 years in Paris, publishing a book on canals and designing a mechanical excavator, a manpowered submarine and a torpedo, all ahead of their time. He believed that these weapons would help in securing universal disarmament. His steamer was successful because he used theoretical knowledge as well as practical skill in its design.

Oliver Evans' high pressure "Grasshopper" engine, so called because of the strange motion of the beam.

Evans' steam dredger. He called it "Orukter Amphibolus". The artist who drew the original picture in 1834 doesn't seem to have understood how it worked.

200 kW side lever engine for the paddle steamer "Dee", built by Henry Maudslay in 1832. Instead of one beam above the cylinder and crankshaft, there are two side levers on either side of the cylinder to reduce the height of the engine.

An early portable engine driving a threshing machine.

The Clermont was 45 m long and used a 20 horsepower (15 kW) Boulton and Watt engine. Instead of an overhead beam, however, the piston drove a pair of side levers through connecting rods and this type of *side lever engine* became an accepted design, saving space and preventing boats from being top heavy.

Oliver Evans made a steam dredger, also in America, in 1804 which he claimed was amphibious but it doesn't seem to have been a great success. It used one of the earliest examples of a *"Grasshopper" engine* in which the beam is pivoted to a lever at one end. This does away with the need for Watt's complicated linkage.

Steam power began to be used on the land too. The Fens were drained, and the Dutch Polders. Trevithick planned a traction engine in 1812, but it was before its time. It wasn't until the 1840s that portable engines, or portable powers as they were sometimes called, were used on the farms in any number. They had wheels so that they could be moved about – and shafts for horses to pull them by. Later, of course, they became self-propelled and were used for reaping, threshing, ploughing and making drains.

This Dutch pumping station was built about 1840. One Cornish engine drove eight pumps; six of the beams can be seen.

Locomotives

Cugnot's steam engine knocked down a wall when it was being demonstrated to the French War Minister.

The development of the Cornish engine meant that factories could be built anywhere, not just by fast-flowing streams and rivers. Steam power soon began to be used for grinding corn, for making cloth and for printing. The Times was first printed on a steam-driven printing press on 29 November 1814.

Another big change came with the building of the railways. It became possible to send raw materials and finished goods about the country easily and cheaply, and for the first time working people were able to travel. Before this, they could only go as far as they were prepared to walk, as the stage coaches which linked the big towns were slow, uncomfortable and very expensive. As early as 1769, Nicholas-Joseph Cugnot, a French army officer, made a steam-driven tractor for towing guns on the road. It was very large, very slow, very hard to steer and would only go a short distance before it had to be refilled with water. Trevithick's light, high pressure engine was much more suitable, and he made two road *locomotives*; the first of these blew up in 1801 when it was left with the fire going while Trevithick and his friends were celebrating with a roast goose dinner in an inn. Steam carriages for use on the roads could have been developed at this time, but many people were against them – the canal owners, the coaching companies and later the railways. They were kept off the roads by charging very high tolls (in those days people had to pay to use the roads) and by passing laws limiting their use. People also said they were dangerous, and they were right; thousands of people

George Stephenson (1781–1848) and his son Robert (1803–1859) formed a famous railway engineering partnership. George was the son of a coal miner, and didn't learn to read or write until he went to night school at the age of 18, but made sure his son had a proper education. George started life as a mechanic; before he died he was offered a knighthood and a seat in Parliament, but refused both. He was the first President of the Institution of Mechanical Engineers. Robert built several famous bridges; he became an MP, and a Fellow of the Royal Society. He died at the early age of 56, worn out by overwork, and was buried in Westminster Abbey.

In 1820, Sir Goldsworthy Gurney's steam carriage regularly carried passengers between Cheltenham and Gloucester.

are now killed on the roads every year. In 1804, Trevithick made a railway "tram-engine" which successfully pulled a load of 10 tons of iron and 70 men at Penydarren ironworks near Merthyr Tydfil, but it was too heavy and broke the rails, which were meant for horsedrawn traffic. He also gave "joy rides" on a circular track in London with an engine

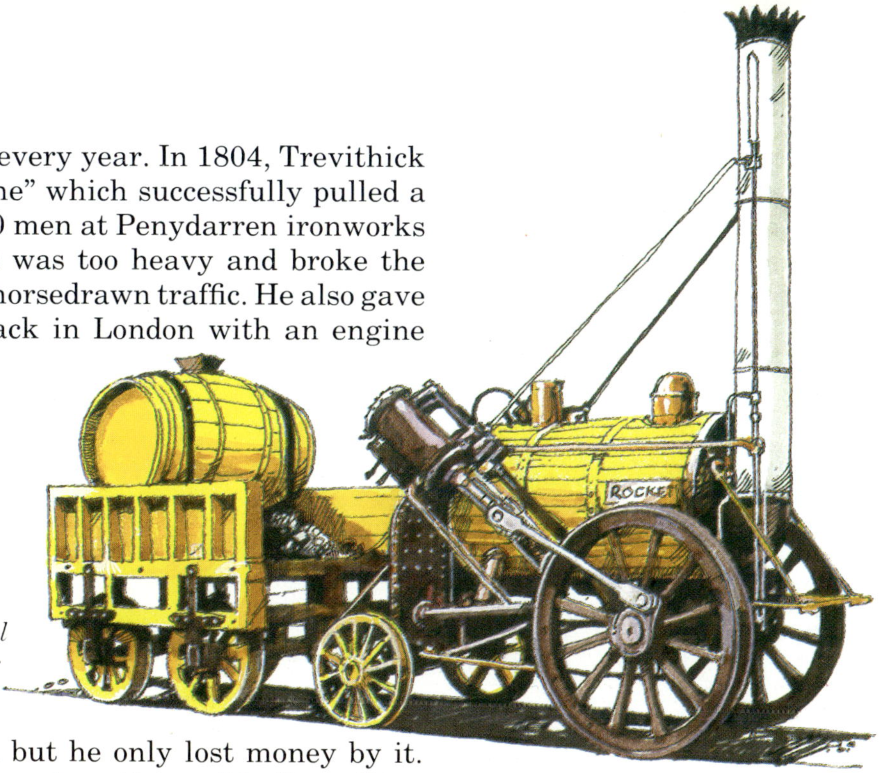

The Stephensons' famous "Rocket". Later the cylinder was lowered to give a more horizontal drive, which resulted in steadier running.

called "Catch me who can", but he only lost money by it. Other people were also experimenting with "travelling engines" for use in coal mines, ironworks and so on.

The first railway that the public could use was the Stockton and Darlington Railway, built by George Stephenson, and opened with great ceremony on 27 September 1825; but the passenger coach was drawn by horses, and so were many of the coal trains. The Liverpool and Manchester Railway was much more important. There were no horses and steam-hauled passenger trains were run for the first time. A "Grand Competition of Locomotives" in 1829 was won by the Rocket designed by George Stephenson's equally famous son Robert. If the Rocket had failed to meet the conditions of the trial, as the other engines did, the railway would probably have used stationary engines and cables to pull the trains. In 1830, the Stephensons supplied the Planet to the Liverpool and Manchester railway. This engine set the pattern for later locomotives, with the cylinders at the front under the *smoke box*, driving wheels at the rear.

The "Planet", only a year after the "Rocket", showed how all future steam locomotives would be designed.

Patrick Stirling's famous "No. 1" of the Great Northern Railway, was designed in 1870. It was one of the most elegant locomotives ever made.

Compound engines

In this later McNaught compound engine the cylinders are on the same side of the beam pivot.

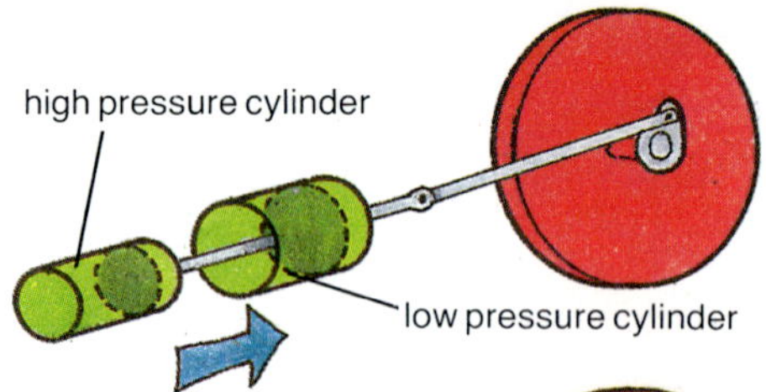

Tandem compound engine.

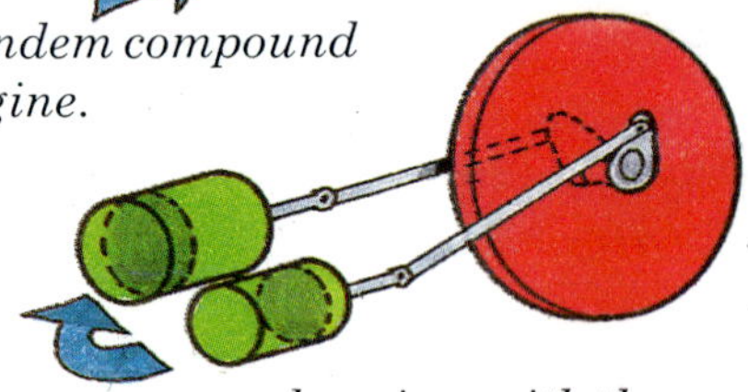

Cross compound engine, with the flywheel between the centre lines of the cylinders.

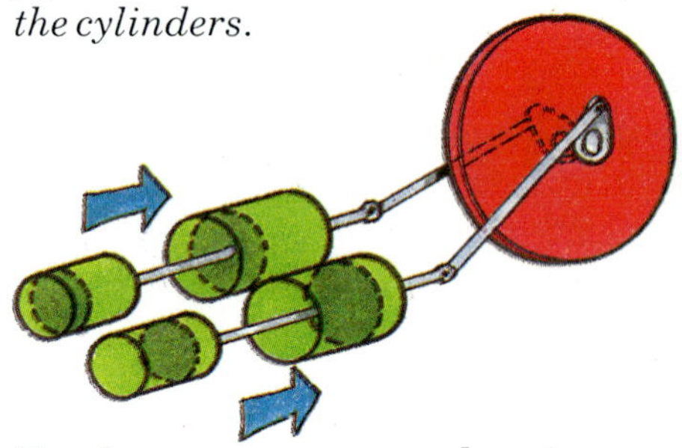

Tandem-cross compound engine.

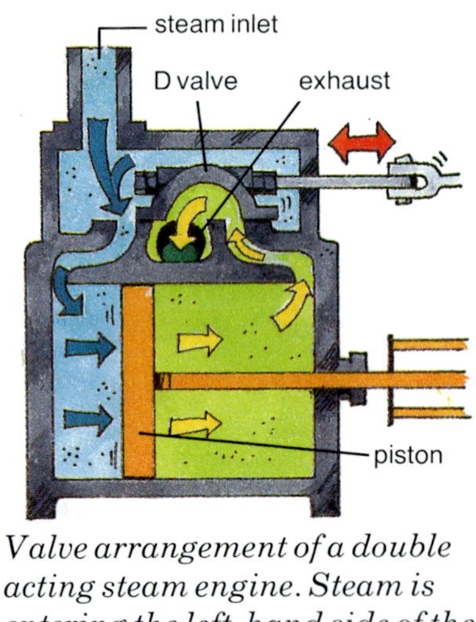

Valve arrangement of a double acting steam engine. Steam is entering the left-hand side of the cylinder and leaving the right-hand side as the piston is driven to the right. After a while, the valve moves to the left and reverses the process.

Using high pressure steam in a single cylinder engine is wasteful. The steam needs to be exhausted at as low a temperature and pressure as possible. A very big cylinder is therefore needed; there is a lot of friction and a lot of heat is wasted in radiation from the cylinder. At the beginning of the stroke, hot steam is coming into a cold cylinder, and this wastes heat too. In a compound engine, the expansion of the steam is split up into stages. The steam goes first into a small, hot, high pressure cylinder and then into a larger, cooler, low pressure cylinder before going to the condenser or to exhaust. The losses are reduced and the engine can be 50% more efficient.

Jonathan Hornblower of Cornwall tried out the idea in 1791, and Arthur Woolf of London in 1803. Neither of their engines was very successful, partly because they didn't know how to calculate the sizes of the cylinders properly, and partly because the steam pressure wasn't high enough anyway – only about 3 bar; three times this pressure is really needed.

William McNaught of Bury near Manchester made a successful compound engine in 1845, by adding a high pressure cylinder to a beam engine on the opposite side to the existing cylinder, between the pivot of the beam and the connecting rod or pump rod. Many beam engines were converted in this way – it was called "McNaughting". Many different types of compound engines followed, some with three and even four cylinders in succession (*triple*, or quadruple, *expansion*).

Boilers

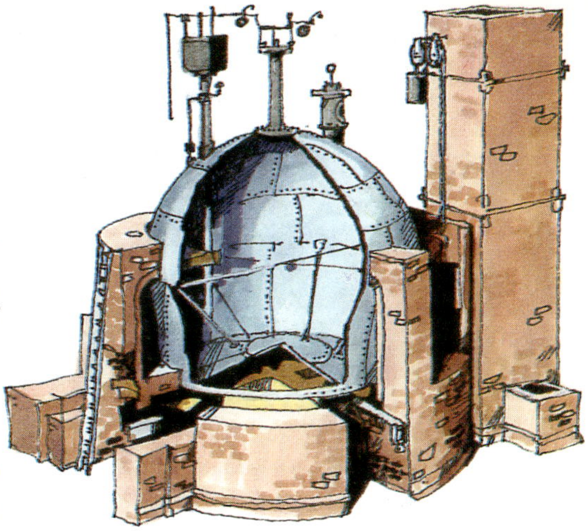

A "haystack" boiler.

Newcomen used a boiler which was just a development of the coppers used in brewing; this was called a "haystack" boiler, because the shape was like that of the circular haystacks of the time. They were made of wrought iron plates, and the fire underneath was contained in a brick enclosure. They could only support a steam pressure a little above atmospheric. Watt produced an elongated version, to produce more steam, called the "Wagon" boiler, the shape being like that of a cart with a hooped cover.

The Cornish boiler had one fire tube. This type was made in Perranarworthal in the 1800s.

Trevithick needed a much stronger boiler for his high pressure steam and in 1812 made a type which came to be called the *Cornish boiler*. It was a horizontal cylinder containing the water, with another horizontal tube running through it to carry the flue gases from the furnace. This increased the area of hot metal in contact with the water.

Later came the *Lancashire boiler* with two flues instead of one. It was developed by Sir William Fairbairn, and thousands of them were in use for many years.

One reason for the success of the Stephensons' Rocket was their use of 25 *fire tubes* to carry the flue gases through the boiler giving a large heating area. Later, locomotives had many more fire tubes than this.

Another important type, the water tube boiler was developed by Herman Babcock and Stephen Wilcox in America in 1861. Here the water was carried in small tubes through the furnace. This type of boiler was often used in ships.

Below left: Lancashire boiler with two fire tubes.

A water tube boiler. The fire is at the left-hand end. Since warm water rises and cold water sinks, the water circulates clockwise through the boiler and the tubes.

The theory of heat

Most of the early work on steam engines was done by practical men, who didn't have much idea of the greatest *efficiency* it was possible to aim at. The study of *Thermodynamics* provided these ideas.

Sadi Carnot (1796–1832), the son of a famous general of the French Revolution, wrote a book when he was only 28 explaining that there is a limit to the efficiency of an engine. The limit depends on the temperatures of the inlet and outlet steam (or other fluid). This idea gives engineers a standard to aim at; but it was some time before it was accepted.

A steam engine uses heat to do mechanical *work*. The maximum amount of work that can be expected from a given amount of heat was determined by an amateur scientist, James Prescott Joule of Manchester (1818–1889), the son of a wealthy brewer.

These ideas were combined and developed by William Thomson, later Lord Kelvin (1824–1907). He was an extraordinary man who qualified to enter Glasgow University at the age of 10, became a Professor there at 22, wrote 700 scientific papers, made over 50 inventions – and laid the first telegraph cable across the Atlantic. However, it was another Glasgow professor, Macquorn Rankine, who brought these very important ideas to the attention of the practical men of the time in 1859 in his book "The Manual of the Steam Engine".

Another important invention arose from these ideas. If a machine can turn heat into work then a machine can also turn work into heat – a refrigerator (or a heat pump, if it is warmth that is required rather than coldness).

William James Macquorn Rankine was born in Edinburgh in 1820. His father was a soldier, who later became an engineer and was in charge of the Caledonian Railway. Young Rankine worked on the railway and as a surveyor in Ireland, becoming a professor at Glasgow University at the age of 35. Apart from his work on heat, he made important studies of the strength of the materials used in engineering, the design of ships, refrigeration and hot air engines. It was largely because of him that Engineering was accepted as a proper subject to be studied at Universities.

A humorous, musical, poetical man, no one seems to have had a bad word to say about him.

Two powerful locomotives of the 1920s. The British one (at the top) looks beautiful, but the German one was much easier to maintain, because the "bits and pieces" were not hidden away.

The steam engine in its prime

The last steam locomotive for British Rail, "Evening Star", was built in 1960. By this time, steam railway locomotives had become more efficient and very powerful – up to 3000 kW – by using *compounding* and *superheated steam* at high temperatures and pressures. They are still used in some countries, but have mostly been replaced by diesel and electric locomotives, which are more efficient still.

Isambard Brunel's "Great Britain" of 1843 was the world's first iron-built, propeller-driven ship. She is now being restored in Bristol. The engine had four cylinders, each one over 2m in diameter, and developed over 1000 kW.

Steam engines for ships caused many problems. Even for a paddle steamer with its high-up drive shafts, the beam engine took up too much space. Many ingenious methods were tried to reduce the height of the engine, such as Maudslay's side lever engine (shown on page 14). With screw propellers, though, the drive shaft is low down, and the standard type of marine engine had its triple or quadruple expansion cylinders placed above the crankshaft.

This now seems to be the right way up for an engine – but Newcomen and Watt would have found the idea impossible.

At first, marine engines used sea water, which corroded the boilers quickly, so condensers were used to recirculate a supply of fresh water.

Many steam lorries were used on the roads for a time. In Britain, until 1896, they had to have a crew of three, with one walking in front. They had to stop every few miles for water. Steam rollers and ploughing engines were also used.

A magnificent Fowler fairground traction engine, built in 1934. It was a three cylinder compound and could be used to drive the powerful electric generator mounted at the front.

Steam turbines

In a piston engine, large forces are needed to accelerate and slow down the piston at the beginning and end of each stroke; so there is a limit to the size and speed of a steam engine – about 15 000 kW and 650 revolutions per minute (*rpm*).

Steam turbines run much more smoothly, so can run much faster. They are more efficient and can develop much higher powers – over 600 000 kW.

Hero's turbine (page 6) is an example of a *reaction turbine*; it works rather like a rocket. In a rocket, the pressure of the burning gas acts on the inside of the front, giving a forward push to drive the rocket. Since there is just a hole at the back for the gases to escape, there is no corresponding backwards force. Branca's (page 6) is an *impulse turbine*, in which the jet of steam hits the blades, and leaves them at reduced speed. The two principles are combined in many modern turbines. Several people realized the advantage of the turbine principle, but none of them was successful in applying it for many years. Richard Trevithick made a "whirling engine" like Hero's, but with two arms each 2½ m long, which revolved at about 300 rpm. It wasn't very successful, but it must have been a fiercesome sight when it was running. In 1891, Carl Gustav de Laval (1845–1913), a Swedish engineer, invented a successful impulse turbine which could drive small machines; but its power was low and it ran at 18 000 rpm – too fast for most purposes. This speed was so much higher than anyone had used before that it caused great problems in design. De Laval was a very ingenious man who made many other inventions.

However, the real inventor of the modern turbine was the Hon. Charles Algernon Parsons – he was an Irish aristocrat, and took the most unusual step for such a person of becoming

Sir Charles Parsons was born in 1854, youngest son of the Earl of Rosse, of Parsonstown in Ireland who was a famous amateur astronomer. Charles studied at home, and then at Cambridge University. There was no course in engineering so he graduated in mathematics. He was an apprentice at the famous firm of W.G. Armstrong in Newcastle for four years. He married in 1883 and in January 1884 started work on his turbine; by April he had patented it. He also made important advances in the design of gear wheels and optical instruments – altogether he took out 300 patents. He was knighted in 1911, and in 1927 was the first engineer to become a member of the Order of Merit. The members are chosen by the Queen (or King) and there are never more than 24 of them at any one time. He died in 1931.

"Turbinia" at speed. Nine propellers were needed to absorb the power of the engine.

An early Parsons steam turbine of 350 kW.

"Turbinia", 1894, 1500 kW steam turbine, and "Mauretania", 1906, 50 000 kW steam turbines.

an engineer; and a very brilliant engineer too. His first reaction turbine produced 7½ kW. It was used to drive an electricity generator, and was completed in 1884. The steam passed through several rows of blades, one after the other, with fixed blades in between. This meant that the speed of steam need not be too high.

Charlie Parsons, as they called him in Newcastle where he worked, was a keen yachtsman. In 1897, he astonished the Royal Navy by appearing at Queen Victoria's Diamond Jubilee Naval Review off Portsmouth in a 30 m yacht, Turbinia, travelling at 34½ *knots* – much faster than the fastest ship the Navy had. He designed Turbinia himself; it was the first boat to be driven by a steam turbine. By 1905 the Navy had decided that all their warships should have turbines. The great ocean liners soon had turbines. The Mauretania, of 44 000 tonnes, was fitted with turbines of over 50 000 kW.

Nowadays, oil burning, coal burning and nuclear power stations all use steam turbines, running at 3000 rpm. A modern power station can burn 20 000 tonnes of coal a day, so efficiency is very important, and condensers are used. Steam is used at temperatures up to 600°C and pressures up to 170 bar; in fact engineers prefer to talk about steam generators rather than boilers – they are very complicated devices, through which water and steam pass several times, and very different from the simple boilers mentioned earlier.

A modern steam turbine. Like steam engines, they are compounded. The high pressure cylinder is nearest, followed by the intermediate and low pressure cylinders. There are thousands of blades altogether.

Hot air engines

A hot air engine built early in the century. It was fired by gas at the bottom.

As early as 1759, it was suggested that air could be used instead of steam as the working fluid in a heat engine; but it was nearly 60 years before the idea was put into practice.

Hot air engines do not need a large water supply; but they have to have twice as many pistons as a steam engine, so they are rather heavy for their output of power. Nevertheless, many were used in farms and small factories until they were replaced by electrical motors and small internal-combustion engines at the beginning of this century. Recently, though, engineers have begun taking an interest in them again. Like steam engines, they are *external combustion engines* – burning of the fuel takes place outside the cylinder. This means that they can run on coal, wood, rubbish, or anything that will burn. So, as oil becomes more scarce, their importance is likely to increase. They also cause much less pollution, are quieter, and can be made much more efficient than internal-combustion engines.

Hot air engines need a *displacer* piston in addition to the power piston which actually drives the crankshaft. Sometimes there is a separate displacer cylinder, sometimes both pistons are in one cylinder. The cranks for the two pistons are at right angles to one another so that when the power piston is at the end of its stroke, the displacer piston is in the middle of its stroke and vice versa.

Robert Stirling was born in Scotland in 1790. He filed his patent for "An Engine Capable of being Applied to the Moving of Machinery on a Principle entirely New" when he was 26. His brother James (1800–1874) probably helped him to put his ideas into practice; even so, Robert still found plenty of time to see to his religious duties, for he was later known as "The Father of the Church of Scotland". He took out further patents for his engines in 1827 and 1845, and also made scientific instruments to his own design. He married in 1819 and had three sons; two became engineers and one became a clergyman. He died in 1878.

How a Stirling engine works. (1) The power piston pushes the air into the cold space. (2) The displacer pushes the air into the hot space. (3) The hot air expands, flows past the displacer, and pushes the power piston to the right on the working stroke. (4) Back to position 1.

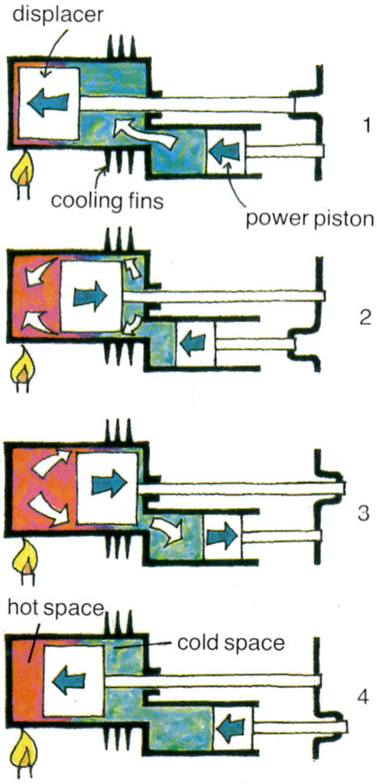

The displacer cylinder has a hot end where the flame is applied and a cold end. The displacer piston is a loose fit so that air can flow from the hot end to the cold end and back again. Thus the displacer piston provides hot air for the working stroke of the power piston, and sends the air back to the hot end to be reheated during the return stroke.

The pioneer of the hot air engine was the Reverend Doctor Robert Stirling, a Scottish clergyman, who worked with his brother James. Hot air engines are now usually called Stirling engines. The first was made in 1816 with a 3 m high cylinder. Another pioneer was Sir George Cayley, a Yorkshire squire who was a brilliantly clever amateur engineer. He was the first to study the theory of flying, built the first man-carrying glider, and invented the caterpillar track and the type of wheel now used in bicycles – where the weight is carried by flexible wires in tension. Many hot air engines were built to his design between 1837 and 1900.

A third important contribution came from John Ericsson, who came to England from Sweden in 1826. More than 3000 engines of his design were made in America. Modern Stirling engines were developed by the Dutch firm of Philips during World War II. Instead of air, hydrogen or helium is often used. These gases, used in airships, are much lighter than air, so less work is wasted in pumping them to and fro in the engine. Research on Stirling engines is going on in several countries, and proposals for using them in cars, railway locomotives and ships are being worked on.

John Ericsson was born in 1803 and served in the Swedish Army for a time. His locomotive "Novelty" competed against the Stephensons' Rocket. Although very fast, it was too light. "Yon thing's got no guts" said George Stephenson, and he was right – it blew up. Ericsson invented the screw propeller for ships. He went to America in 1839, where he designed an ironclad warship, "Monitor", with a huge gun – the first to use a revolving gun turret, and the first without sails. Later he became interested in more peaceful things, particularly in the idea of producing energy from the sun, the wind, the tides, and the earth's gravity. At one time he was making a major invention every day. He worked up to 14 hours a day, right up to his death at the age of 86 in 1889.

A Philips portable electric generator of 1950 weighing 30 kg, powered by a hot air engine of modern design. The displacer is above the power piston in the single vertical cylinder on the left.

Gas engines

Many people thought about the idea of the internal-combustion engine – burning or exploding something in a cylinder to drive a piston. An engine which didn't need a boiler would be much more convenient than a steam engine - the big ocean liners like the Mauretania needed hundreds of men just to shovel coal into the boilers. This was hot, hard and very unpleasant work. The first idea came from Christian Huygens, Denis Papin's employer, who suggested using gunpowder to produce a vacuum under the piston of an *atmospheric engine* – but a fluid like steam was obviously much more convenient than this. Other people suggested using hydrogen, alcohol and turpentine vapour.

The real answer, though, was gas. This was first produced in a commercial gasworks for use in lighting in 1812, but it wasn't until 1860 that it was successfully used in an engine by Etienne Lenoir in France. Lenoir's engine was just like a horizontal double acting steam engine – it even used the same sort of slide valves. Gas was let into either end of the cylinder part way through the stroke, and set alight by an electric spark. In fact, Lenoir's engine wasn't as efficient as the steam engines of the time, but it was small and convenient. Over 400 were made, giving up to 2 kW – the first successful internal-combustion engines.

To be really efficient, the gas needs to be compressed before it is set alight, and afterwards it needs to expand as much as possible. A Frenchman, Alphonse Beau de Rochas (1815–1893) realized this in 1862, and patented the idea of the *four-stroke engine* (used in most cars today). He didn't actually make an engine; this was done by Nikolaus Otto,

Etienne Lenoir was born in Luxembourg in 1822, but moved to France that year. He taught himself engineering, and made many inventions apart from his gas engine – in railway signalling, electric motors, and sending pictures by telegraph. He used his gas engine, rather unsuccessfully, in a carriage and in a boat – the first time internal-combustion engines had been used like this. He was not a very good businessman and lost his money when Otto and Langen's engines began to be sold; he was poor when he died in 1900.

Opposite: The four-stroke Otto cycle. (1) Induction. Gas and air are sucked into the cylinder. (2) Compression of the gas-air mixture. (3) Power stroke. The mixture burns and drives the piston down. (4) Exhaust. The cylinder is emptied ready for the cycle to begin again.

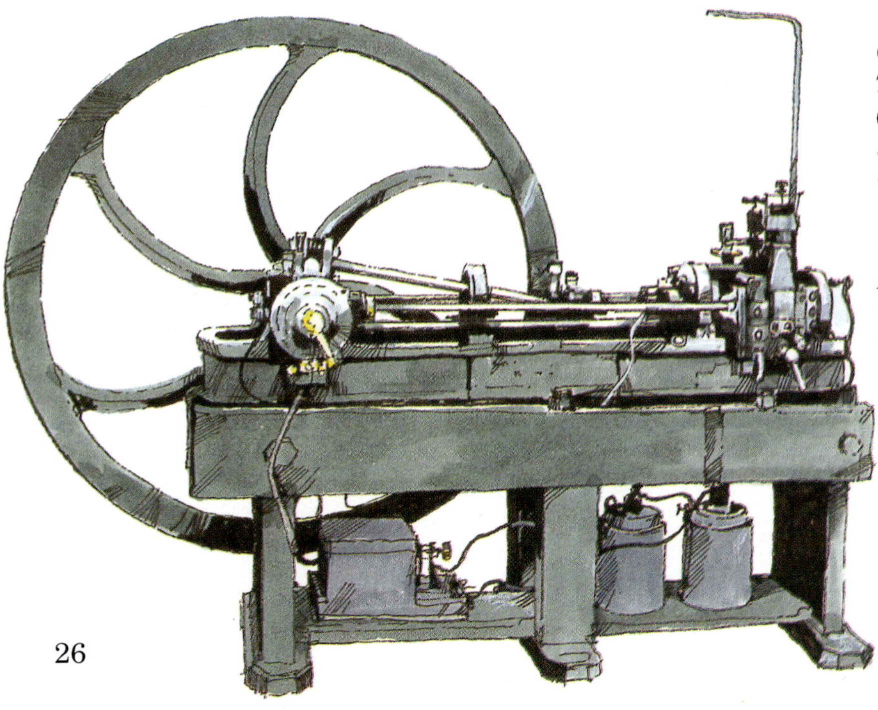

Lenoir gas engine built at Reading in the early 1860s.

Otto's first experimental four-stroke gas engine of 1876.

Nikolaus Otto was also self taught. He was the son of a German innkeeper and farmer and was born in Nassau in 1832. He worked as a clerk until he read one day in a newspaper about Lenoir's engine and decided to make engines himself. In 1864 he met Eugen Langen and the two started to work together. The firm employed two men, Maybach and Daimler, who later became very famous for their work on engines and on motor cars. Otto was a modest, retiring sort of man, even after his success. He died in 1891.

a German engineer, who went into partnership with Eugen Langen (1833–1895), an engineer and businessman who was working in his family's sugar refinery. Their first engine was a gas-driven, atmospheric engine, but in 1876, they produced the first four-stroke engine. It produced 2¼ kW at 180 rpm, twice as fast as Lenoir's engine, and much more efficient. It became known as "The Otto Silent gas-engine" – though today it would be considered pretty noisy – and tens of thousands were built at Otto and Langen's "Gasmotoren-Fabrik" at Deutz in Cologne and in many other places as well. Otto wasn't able to patent the idea, because Beau de Rochas had already done so. The four-stroke *cycle* on which it worked became known as the Otto cycle.

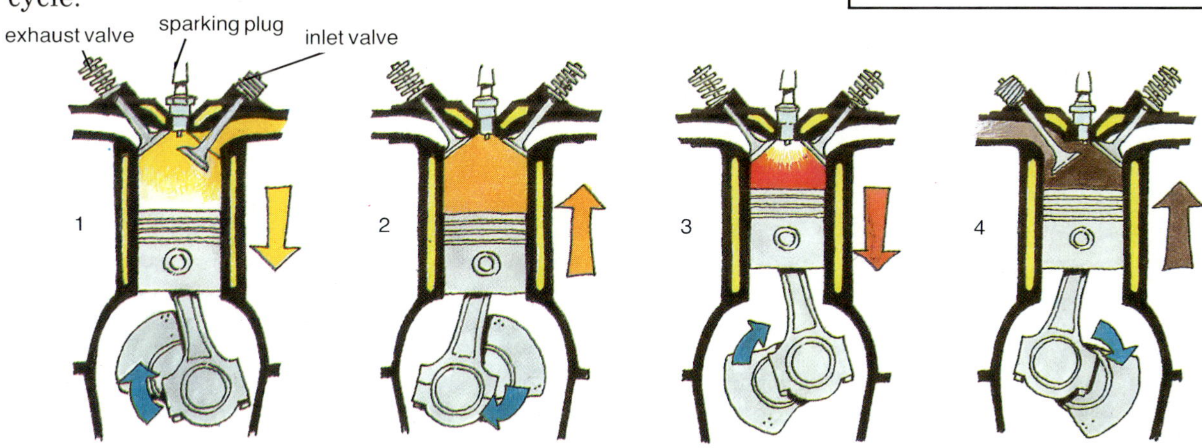

Most gas engines were run off the town's gas supply. In 1881, Joseph Dowson (1844–1940) developed a gas producer which made gas suitable for driving an engine from coal or coke, so that gas engines could be used anywhere. Gas engines giving as much as 4000 kW were eventually built.

Petrol engines

Karl Benz, born at Karlsruhe in 1844, soon became an orphan – his father was the first engine driver to be killed in a railway accident. He became a *fitter* at the Karlsruhe steam engine works, but after he married in 1872, he set up his own firm at Mannheim, the first one ever to sell motor vehicles. He sold his first car, a three-wheeler, in 1887, and by 1888 had 50 men working for him. A four-wheeler followed in 1893. He had to get permission to try his car out on the road, because it was against the law. In 1885 his wife, Berthe, an adventurous lady, made the first long-distance car drive ever, to her home 100 km away, without telling Karl, but with the help of her two teenage sons. They had to make several stops for repairs, including a visit to a blacksmith – there were no garages, of course. Benz died in 1929.

Above left:
Benz's motor tricycle of 1885 had a single horizontal cylinder with a vertical crankshaft. The huge flywheel can be seen at the back.

The petrol engine is the type that people are most familiar with, because it is used in cars and motorcycles. In fact, it made the car possible and the aeroplane as well. Neither the steam engine nor the gas engine were suitable (although there were a few steam carriages early in the 19th century, and some very good steam cars early in this century).

The petrol engine is light, because it doesn't have to carry a boiler or a gas producer around. It can start instantly, unlike a steam engine, and can run for many hours at high speed without any attention – and it is much cheaper to run than relays of horses. It wasn't until 1859, though, when the first oil well was drilled (see "Resources" in this series) that the idea of the petrol engine became practical.

The first were made independently by Karl Benz and Gottlieb Daimler in Germany in 1885. Benz had the better idea, really; he wanted to design a vehicle to replace the horse-drawn carriage. Daimler at first just wanted to replace the horse, and got a coach-maker to build him an ordinary carriage without shafts to try out his engine. However, Daimler's first engines were better than Benz's. They were regarded as "high speed" engines, because they would go at 800 or 900 rpm, much faster than any other engines of the time. Nowadays, ordinary car engines will run at 5 or 6 times this speed, and a motorcycle engine has been made to run at 24 000 rpm. Similarly, aero engines have been made to produce 4000 kW compared with less than 10 kW for the petrol engines of the 1890s.

Before these increases in performance could be made, many problems had to be solved; how to turn the liquid petrol into vapour, how to get the right mixture of air and petrol vapour into the cylinder – and out again, after it had burned – how to drive the valves that did this, how to get the mixture to burn properly at the right time, how to keep the

Daimler's 1886 "Horseless Carriage".

Gottlieb Daimler (1834–1900) studied mechanical engineering and joined Nikolaus Otto's firm in 1882, where he designed a petrol engine. In 1882, Daimler set up his own factory at Cannstatt near Stuttgart. He first tried his own enginc in a motorcycle with wooden wheels. In the first "Race for Horseless Carriages" from Paris to Rouen in 1894, the 15 cars that finished all had Daimler engines. A new Daimler model of 1901 was called Mercedes, after the daughter of Emil Jellinek. He never met his arch rival Benz, but their firms joined to form Mercedes-Benz in 1926.

engine cool, to make it run smoothly, and make it strong enough.

The usual way of getting the mixture of fuel and air into and out of the cylinder is by means of mushroom-shaped poppet valves. These are usually operated by a *camshaft* and push rods, and closed by a spring. In a four-stroke engine, the valves only operate once in every two revolutions of the crankshaft, so the camshaft has to run at half the speed of the crankshaft.

The drawings show some common valve arrangements. Side valves were simple, but not very efficient, and overhead valves replaced them. To drive these, rockers are often used, but an overhead camshaft is better. Driving the camshaft itself is then the problem; trains of gears from the crankshaft are expensive, and chains are noisy. Nowadays, pulleys and toothed rubber belts are often used. They are reinforced with steel so that they will not stretch.

Different valve arrangements:

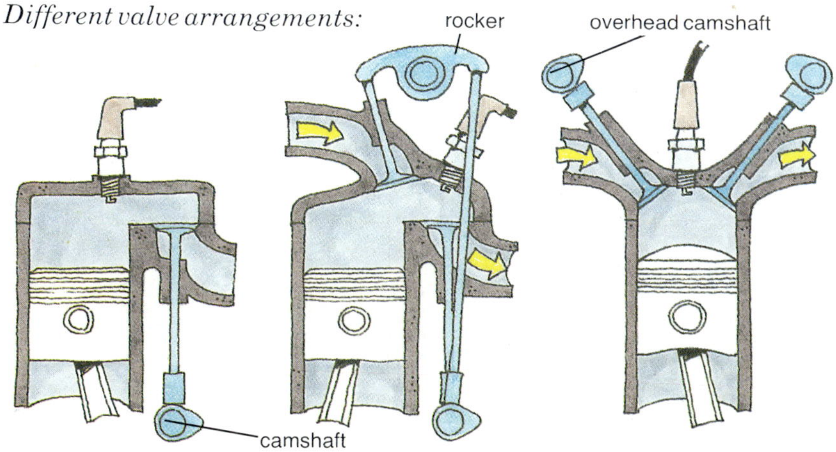

Side valves. *Inlet over exhaust.* *Twin overhead camshafts.*

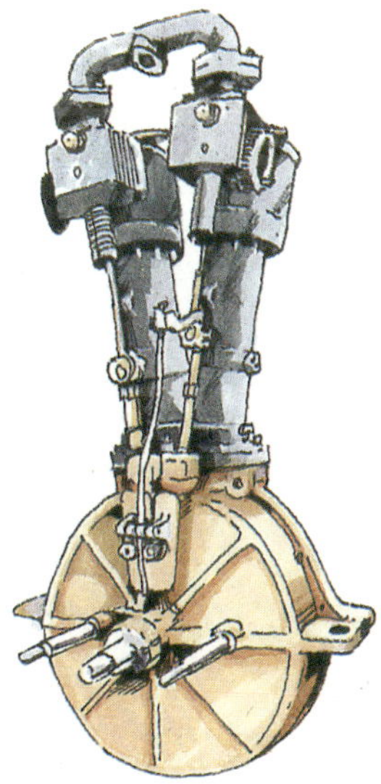

Daimler's twin cylinder V-engine of 1889.

For use in a motor car, a petrol engine has to work under widely varying conditions of load and speed, so the flow of fuel, and the timing of the spark that sets fire to it, have to be carefully controlled. In the cars of the 1920s, the driver himself had to do a lot of the work – he had two levers on the steering wheel controlling the fuel and the spark, and he had to juggle with these continually as he drove along. Perhaps it was fortunate that there were not many other cars on the road to keep a look out for. Nowadays the *carburettor* and the *ignition system* do these jobs automatically.

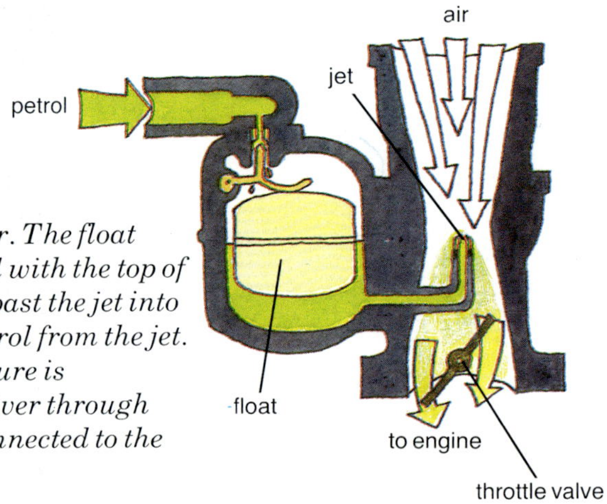

A simple carburettor. The float keeps the petrol level with the top of the jet. Air flowing past the jet into the engine sucks petrol from the jet. The flow of the mixture is controlled by the driver through the throttle valve connected to the accelerator pedal.

The first carburettor of the modern type was designed by Wilhelm Maybach, who worked for Gottlieb Daimler, in 1893. Previous types had worked by allowing the air to bubble through the petrol, but Maybach used the flow of air past a jet in which the level of the petrol was kept constant by a float. As the engine speeds up, the flow of air increases, so more petrol vapour is drawn up with it. A modern carburettor is a complicated device, with extra jets to provide extra supplies of petrol for starting, for slow running, and for sudden acceleration. Another type of carburettor has only one jet; the size of it is varied by a tapered needle down the middle, which is raised by a piston operated by the air flow into the engine.

Igniting the mixture in the cylinder is a difficult problem. The timing has to be just right or the engine will run roughly, noisily, and inefficiently; and at a speed of 6000 rpm, the whole of the power stroke only occupies $\frac{1}{200}$ of a second. Some people used naked flames. Others used *hot tubes* projecting into the cylinder and heated by naked flames outside the cylinder. These methods were only suitable for slow-running engines, and were quite dangerous as well. Striking a spark from a flint was also tried, but the electric spark proved to be much the most suitable method. It was used by both Lenoir and Benz.

Wilhelm Maybach was born in 1847. Like Benz, he was an orphan, and had a hard childhood. He joined Otto's gas engine works in 1812, at the same time as Daimler, and went with him when Daimler formed his own firm. He was an engineer of genius, and helped Daimler with many of his design problems. In eight years, until Daimler's death in 1900, they produced several different engine designs between them. It was Maybach who, with Daimler's son Paul, designed the first Mercedes car of 1901, setting the pattern for the future. In 1901 he joined Count Zeppelin (1838–1917), the pioneer builder of airships. Maybach, helped by his son Karl, designed many engines for airships, aeroplanes, buses and lorries. He died in 1929.

Ignition and mixture controls on the steering wheel of a car of the 1920s.

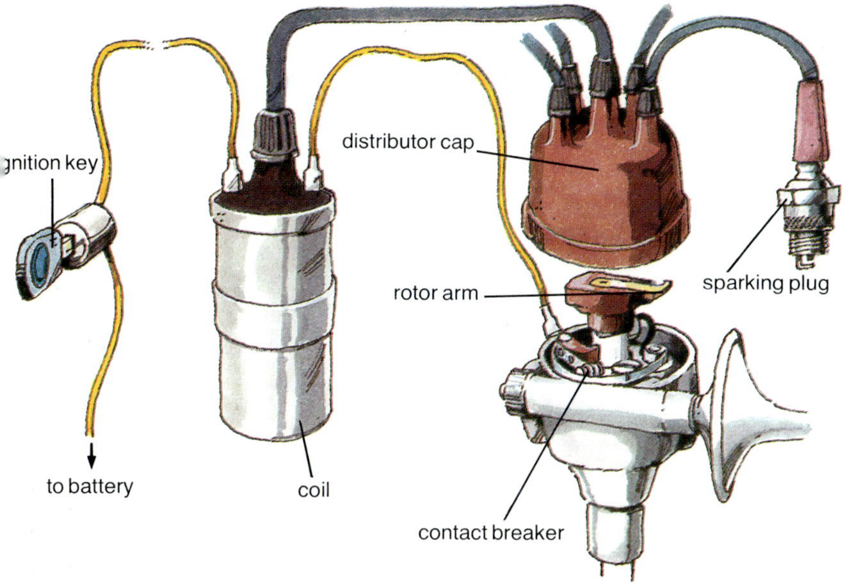

Coil ignition system. The coil transforms 12 volts from the battery to several thousand volts at the sparking plug in the cylinder when the contact breaker opens. The rotor arms send this voltage to each sparking plug in turn.

The spark on most motor car engines is provided by means of a battery, coil and contact breaker. Aircraft engines use small electrical generators called *magnetos*, which do not have to rely on batteries; and they usually have two, so that if one goes wrong the engine still keeps running. Magnetos were developed by a German electrical engineer, Robert Bosch (1861–1942).

Many aircraft and motorcycle engines are air-cooled; many thin fins are machined on the cylinders to increase the cooling effect of the air flowing past them. So were the first motor car engines; but most now are cooled by water – or rather, water mixed with chemicals to prevent freezing and reduce corrosion. Henry Ford (1863–1947) was the first to use water cooling in 1896. The cooling water is itself cooled by air flow through the radiator at the front of the car, assisted by a fan. Fans are sometimes electrically driven and switch themselves on only when required, to save power.

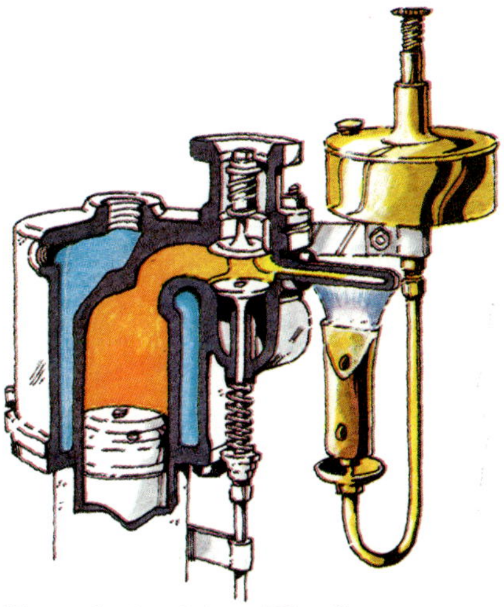

Hot tube ignition. The flame was fuelled from the brass tank at the top.

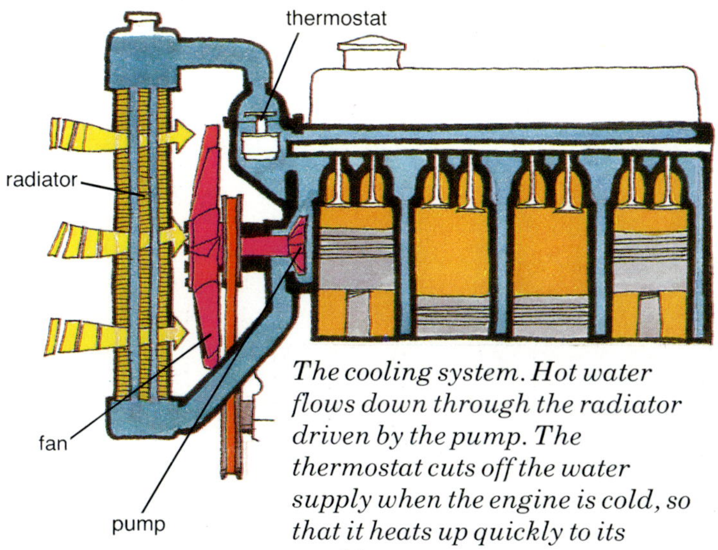

The cooling system. Hot water flows down through the radiator driven by the pump. The thermostat cuts off the water supply when the engine is cold, so that it heats up quickly to its working temperature.

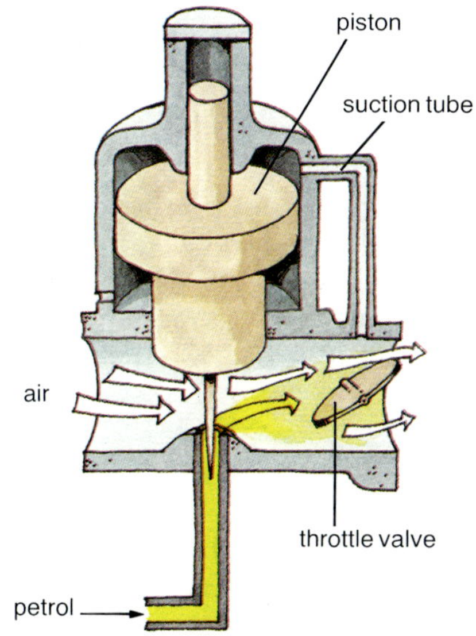

S.U. carburettors were fitted to many sports cars. Air flow past the end of the suction tube gave a partial vacuum above the piston which was raised by atmospheric pressure underneath. This allowed more petrol to flow from the jet.

Diesels

Rudolf Diesel was born in Paris in 1858. His parents had to move to London in 1870 because of the war between France and Germany, but sent Rudolf at the age of 12 to Germany to stay with his uncle. While he was working as a refrigeration engineer, in Paris and in Berlin, he developed his ideas; his patent for a "universal economic engine" was granted in 1897, and later he published a book on the subject. Some years of difficult experiments followed – high pressure can be dangerous, and Diesel was nearly killed in an explosion. Diesel was a pacifist, and hoped that the use of his engine would make life easier for working people. He soon became a millionaire, but couldn't manage his money. He was on his way to England for a conference in 1913 when he disappeared from the Channel steamer, leaving only £1000. Whether or not he committed suicide no one knows.

Petrol has three disadvantages; it will easily explode at ordinary temperatures, it is expensive to refine from the crude oil that comes from the oil well, and it forms less than a fifth of this crude oil. The first engine to use heavy oil was made by George Brayton (1830–1892) of Boston in America in 1874 – before the first successful petrol engine. It had an air compressor to spray the fuel into the cylinder (*blast injection*) where it was set alight by means of a flame. Many of Brayton's "Ready Motors" were made, but they used a great deal of fuel.

Three Quaker brothers of Hull, William, Samuel, and Albert Priestman developed an oil engine, similar to Lenoir's gas engine in 1886, using spark ignition; they were three times as efficient as Brayton's engines. The Priestman engine was greatly improved by Herbert Akroyd Stuart

Above left:
Brayton's motor would run on oil or gas. Air compressed in the pump was stored in the tank ready for use in the working cylinder. Piston, pump and crankshaft were connected to the beam.

A Priestman engine. They were rather difficult to start, but this one was still working during World War II.

This Akroyd Stuart engine of 1890 was cut open as a demonstration model.

(1864–1927) in his father's factory at Bletchley. Instead of using blast injection, the fuel was squirted into the cylinder in carefully timed jets by means of a high pressure pump (solid injection – the method always used nowadays). It was ignited by the heat of the cylinder walls, which had to be heated up before the engine would start. Made by Richard Hornsby & Sons of Grantham, a great many Hornsby-Akroyd engines were produced.

The man who gave his name to the modern oil engine was a German, Rudolf Diesel. A brilliant engineering student, he was shocked when he learned in a lecture that less than 10% of the fuel used in an internal-combustion engine was actually put to good use. He determined to design an engine in accordance with the theoretical ideas of Sadi Carnot. His idea was to compress the air in the cylinder to a very high pressure, up to 35 bar, to get the greatest efficiency. He used blast injection, requiring a higher pressure still to blow the fuel into the cylinder. He found that the temperature of the air already in the cylinder was so high that the fuel set itself alight, and no separate ignition system was needed. His engine of 1897 was so successful that it had an efficiency of over 25%, instead of less than 10%. Because of the high pressure, diesel engines have to be strong and heavy; they are expensive to make and rather noisy, so they are not often used in cars. However, they are much more reliable, longer-lasting and more efficient than petrol engines, and are used in factories, ships, lorries and locomotives.

Large, modern diesels are usually *two-stroke*, not four-stroke engines; each cylinder gives a power stroke in every revolution, instead of in every other. Sir Dugald Clerk (1854–1932) first tried out the two-stroke cycle in a gas engine in 1878. The power and exhaust strokes are combined into one, and so are the induction and compression strokes. This leads to some loss of efficiency, so a two-stroke is not twice as powerful as a similar four-stroke; but it is much simpler, because it needs no valve gear, and runs more smoothly. Many small petrol engines, in motorcycles and lawn mowers, are two-strokes.

Diesel's second experimental engine first ran in January 1894.

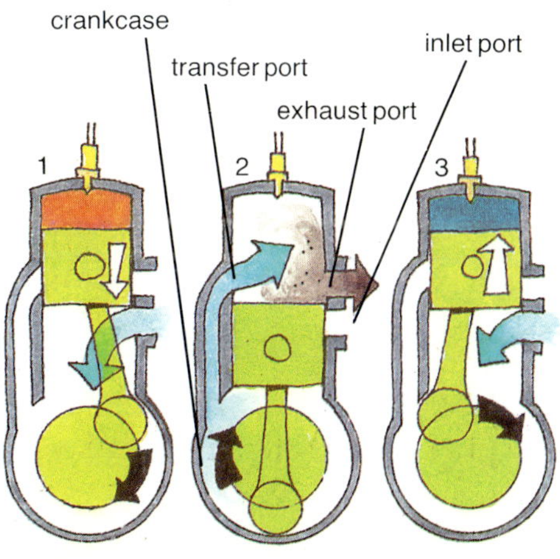

The two-stroke cycle. (1) Power stroke; some air is sucked into the crankcase and compressed. (2) Air from the crankcase pushes the exhaust gases out. (3) Clean air is compressed ready for the next power stroke, while the next charge of air is sucked into the crankcase.

Modern diesel engines

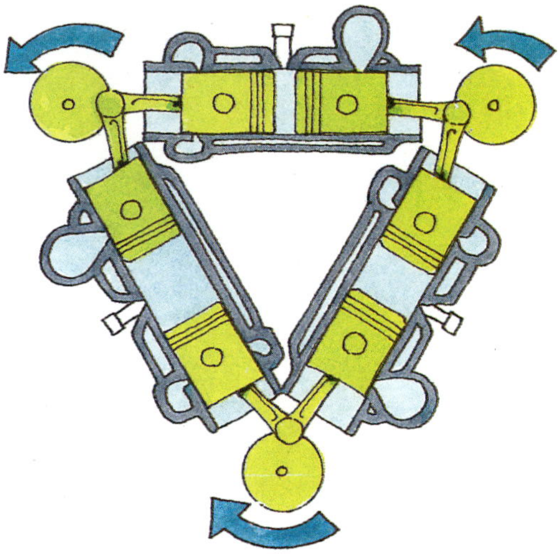

Napier Deltic engine.

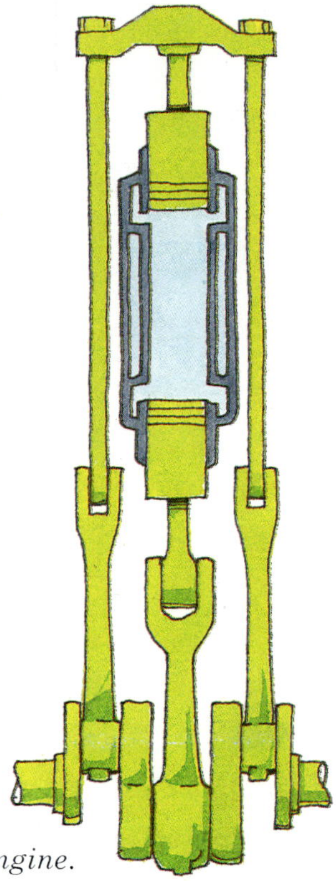

Doxford opposed piston engine.

Diesel engines soon grew in size, and marine engines have been made to produce as much as 36 000 kW. The first diesel-engined sea going ship was made in Russia in 1904; now all but the very largest ships use diesel engines rather than steam turbines. The diagram shows an interesting design, the Doxford *opposed piston engine*, which has no *cylinder head*; combustion takes place in the space between the two pistons. A development of this is the Napier Deltic engine, which has eighteen cylinders and thirty-six pistons arranged in a triangle; there are three crankshafts, with the bottom one rotating in the opposite direction to the two upper ones. Diesels of this sort have been used in locomotives, as stationary engines and in small ships.

The first diesel locomotives went into service in Denmark in 1929. Most modern locomotives have diesel engines, usually generating electricity to drive electric motors which in turn drive the wheels. This works better than using a gear box as in a car, because the power required is so great.

Lorries and buses need smaller, high speed engines and the development of these high speed diesels was largely due to an English engineer, Sir Harry Ricardo. His famous book "The High Speed Internal-Combustion Engine" was first published in 1923. At that time few diesels could exceed 600 rpm; by 1940, 3000 rpm was common. These units were usually four-stroke, very like petrol engines in their layout, and with their development it became possible to make economical tractors and earth moving machines as well.

Harry Ralph Ricardo was born in 1885. His mother came from a famous civil engineering family, and his father was an architect. Even at school he was designing and making engines. One of these became very popular with the fishermen of Shoreham by Sea, where he set up his own firm, to design engines and to advise other people who had design problems. Much of his work was concerned with the shape of the *combustion chamber* between the cylinder head and the top of the piston; it seems a small thing, but it is vital in the efficient working of all kinds of internal-combustion engines.

He was a happy man, fond of sailing and collecting butterflies, and when he died in 1974 he left a large family of grandchildren and great grandchildren.

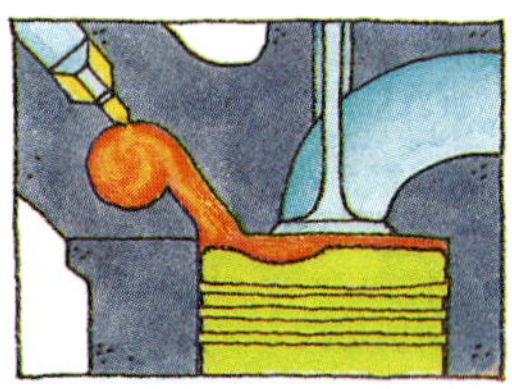

One of Harry Ricardo's combustion chamber designs. It gave a swirling motion to the air and the fuel, resulting in better combustion.

Aero engines

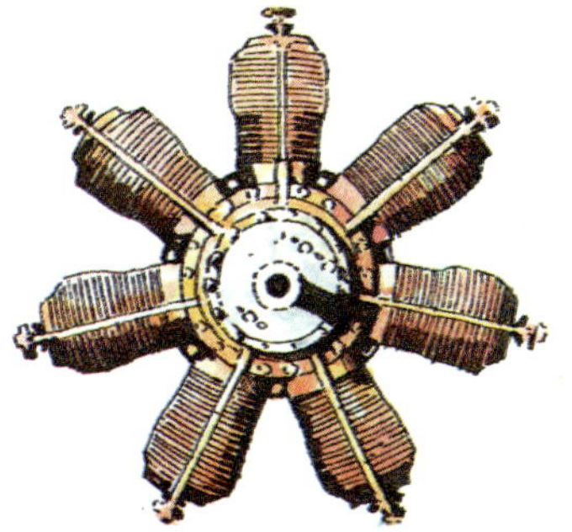

The Wright brothers had to design and make their own petrol engine before they succeeded in making the first powered flights in 1903. Light aircraft engines of today, of the same size and weight, produce eight times as much power at 2½ times the speed – and are much more reliable!

Many of the aero engines of World War I were rotaries – a strange design in which the crankshaft stood still while the pistons, cylinders and crankcase rotated around it, with the propeller bolted on. The first was the famous Gnome of 1907 – the first really reliable aero engine. Rotary engines made flying difficult because they acted as big gyroscopes, and when the pilot started to dive or climb, the aircraft would veer to one side or the other, and vice versa.

Nearly all aero engines use petrol, but airships used diesel engines for safety, and a few aeroplanes were fitted with diesels between the wars.

In World War II, the most famous aero engine was the Rolls Royce Merlin, a liquid cooled V-12, which powered the Spitfire and the Hurricane fighters.

The Gnome rotary engine, used in many of the fighter aircraft of World War I.

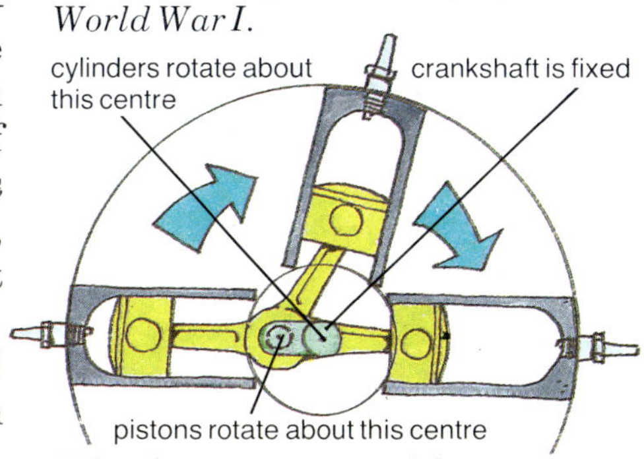

What happens to one of the cylinders and pistons of a rotary engine as it turns through 180°.

Rolls Royce Merlin engine. First designed in 1932, it was based on the engine of the S6b shown on page 4. By the end of World War II, the engine was producing 1500 kW.

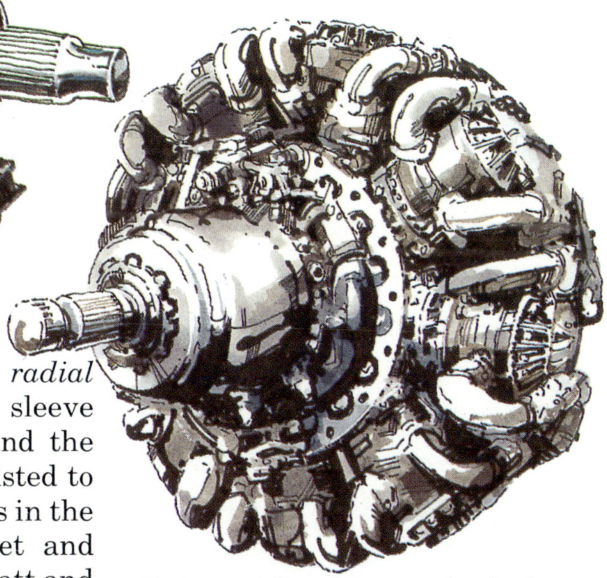

Bristol Hercules and Centaurus were air-cooled *radial* (not rotary) engines, unusual because they had sleeve valves. A cylindrical sleeve between the piston and the cylinder was driven by a small crank, so that it twisted to and fro and went up and down. Suitably shaped holes in the sleeve and the cylinder itself controlled the inlet and exhaust of the gases. The Wright Cyclone and the Pratt and Whitney Wasp were similar American engines, but with poppet valves.

In order to fly at great heights, where the air is thin, piston engines need to have air forced into them by an air compressor, called a *supercharger*.

The Bristol Centaurus engine had 18 cylinders in two rows. It produced well over 2000 kW and was very reliable. It had a much bigger frontal area than the Merlin, however.

Gas turbines

Piston engines, whether two-stroke or four-stroke, produce their power very unevenly; an engine which works steadily and continuously has obvious advantages. The gas turbine is such an engine. It is rather like a steam turbine, but liquid fuel is burned inside it to produce the gas which operates it.

In a gas turbine, a fan-like compressor draws air into the engine and compresses it. Then fuel is burned in the compressed air in a combustion chamber. This causes the air to expand again, and the stream of air is used to turn a turbine. The turbine is connected to the compressor to drive it, and can also be used to drive a propeller for a ship or aircraft, or a generator to provide electric power for a locomotive.

A number of different attempts to design and build gas turbine engines were made between 1872 and the 1930s, none of them very successful, since they were usually based on steam turbine ideas. Success came when two people realised independently that the gas turbine was ideal for use in high speed, high flying aircraft, as a *jet engine.* In this the exhaust gases flow out of a nozzle at the back providing a forward thrust, in the same way as a rocket (see page 22). These people were Frank Whittle, an RAF officer, and Hans von Ohain, a German engineer. The first jet-propelled aircraft to fly, fitted with Ohain's engine, was in the Heinkel He 178 in 1939; while Whittle's engine first flew in the Gloster E 28/39 in 1941. The first squadrons of jet fighters, British and German, went into action in 1944; but the German machines proved unreliable and dangerous, and 27 out of 30 in the first squadron were lost in a month – few of them in action. It was Whittle's engine which proved to be the type on which the development of jet engines after the war was based at first.

Sir Frank Whittle was born in 1907 in Coventry. His father was a mechanic, and Whittle entered the RAF as an engineering apprentice when he was 16. He was one of the very few apprentices to be accepted for pilot training; and became an instructor and a test pilot. His first patent for a jet engine was granted in 1930, when he was only 23. While studying at Cambridge he started his firm, Power Jets, to build an engine. He had to use private money, because the government would not lend him any, until the war began. Whittle retired from the RAF as an Air Commodore in 1948 and received many honours, including a knighthood. Power Jets was taken over by the Government, and Sir Frank now lives in America.

Whittle's W1 jet engine, on which work began in July 1939.

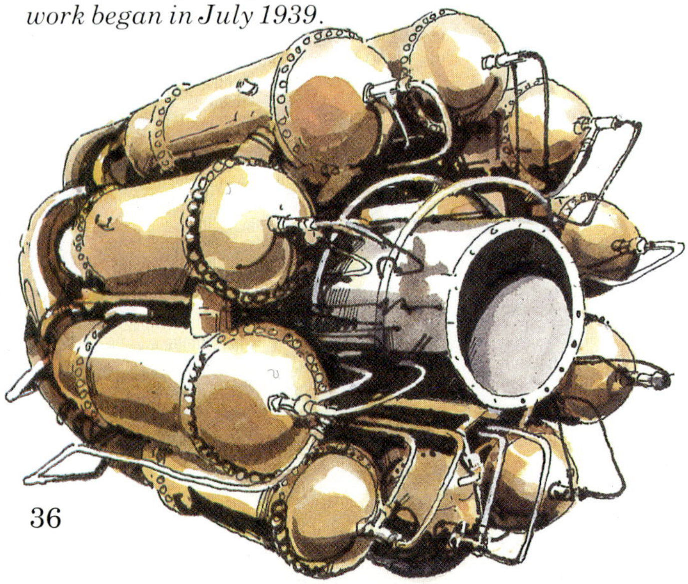

An early type of jet engine with the air flowing outwards through the compressor. In modern engines, the air flow is straighter.

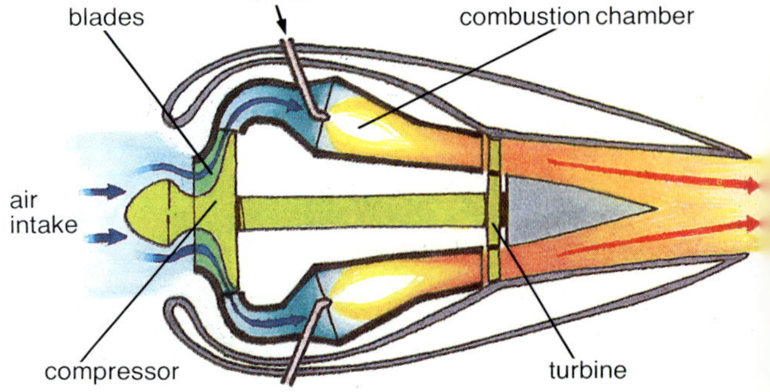

The performance of an aircraft jet engine is not measured in kilowatts of power, but as the thrust force it can exert on the aeroplane. Whittle's first engine ran at 16 500 rpm and gave a thrust of 4½ kN, about the same as a Merlin engined Spitfire – but the performance didn't decrease as the aircraft got higher. Modern aircraft engines give thrusts of 50 times as much; and can run for 10 000 hours without needing to be overhauled. Without the jet engine, the supersonic Concorde would not be possible, nor would the jumbo jets. Because of these, many people can now go on holidays abroad, instead of only the very rich, as in the past.

The Gloster E28/39 experimental aircraft, fitted with Whittle's W1 engine, first flew in May 1941.

The direction of the thrust of a jet engine can be changed fairly easily; in many aircraft the jet is deflected forward when the plane is landing to act as a powerful brake. In the Harrier fighter the jet nozzles can be swivelled downwards to lift the plane off the ground vertically if there isn't a runway available.

Gas turbines are also used in small, fast warships and in industry; North Sea oil rigs have gas turbine engines to supply the power required – and the exhaust gases are put to good use in providing heating. They are also used in power stations, particularly to meet sudden unexpected loads, since a gas turbine generator can be started in 5–15 minutes, compared with 8–12 hours for a steam turbine.

The Rolls Royce Spey engine of 1960 is expected to continue in use for another 20 years; thousands of them have been built.

Rocket motors

Oxygen is needed to burn fuel, and most engines get their supplies from the Earth's atmosphere; but there is no oxygen in outer space, and engines for space craft have to carry their own with them. At ordinary temperatures, oxygen is a gas and takes up much too much room, so it has to be cooled down to a temperature of about −200°C when it is liquid.

Solid fuel rockets were used as weapons of war as early as the 13th century, first of all in China. In 1760 the Indian prince Hyder Ali of Mysore and his son Tipu Sahib used thousands of rockets, of about 4 kg, with a range of 2½ km, against British troops in India. A British artillery officer, Sir William Congreve, spent years during the Napoleonic Wars experimenting with rockets and rocket batteries were used in the battle of Leipzig in 1813. 40 000 of them, 15 kg each, were fired at the city of Copenhagen in 1807. They were also used by the British during the American War of Independence, in the 18th century, and by American troops in Mexico in the 19th century – and during World War II. However, they were never as accurate as guns, which for a long time continued to be the best way of killing people.

Sir William Congreve (1772–1828) with one of his rocket guns.

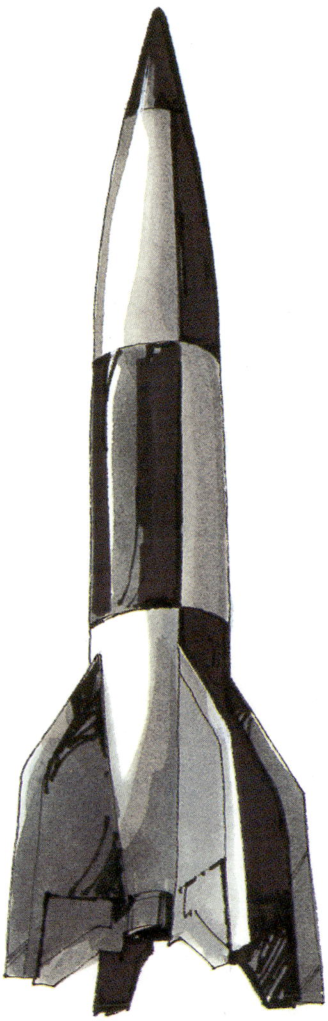

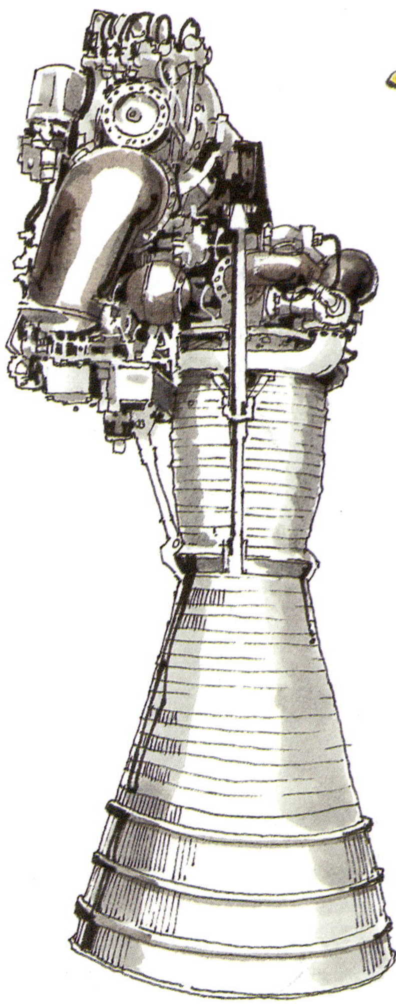

Far left:
The German V2 rocket was 14m long. Over 1000 of them fell on England, killing nearly 3000 people, until the Allied armies captured the launching sites in France, Belgium and Holland.

This Rolls Royce rocket motor gave a thrust of nearly 700 kN.

The German V 2 rockets of World War II were the first really successful liquid fuelled rockets; the fuel was alcohol. They were built by slave labour, and over a thousand fell on London in 1944 and 1945. Fitted with their own navigation system, 8½ tonnes of fuel lifted 3 tonnes of rocket plus 1 tonne of explosive to a height of 100 km at a speed of 3600 mph (1600 metres per second). At this *supersonic* speed there was no warning of their approach (so no one could take cover) and no means of stopping them, and they did dreadful damage. Wernher von Braun (1912–1974), their designer and a member of the Nazi party, took his design team to America after the war to start the American space programme.

Robert Goddard (1882–1945)

Wernher von Braun (1912–77).

An American professor of physics, Robert Goddard, was also a pioneer of rocket engines for space travel; but he got no support from the American government, and worked alone for years. In 1919 he wrote a book "A Method of Reaching Extreme Altitudes", and by 1935 had launched rockets which went faster than the speed of sound. He died shortly before the end of the war.

Even before this, though, a Russian schoolteacher, Konstantin Tsiolkovsky (1857–1935), had discussed most of the problems of space travel in his book "Exploration of Cosmic Space by means of Reaction Devices", written between 1896 and 1903. He was much too poor to make any experiments; nevertheless it was the Russians who first put a satellite, the 83 kg Sputnik, into orbit in 1957, and a man into space – Yuri Gagarin, in 1961.

The principle of a rocket engine is very simple; fuel, usually kerosene, and oxygen are mixed together and set alight. But, because of the danger of explosions, the low temperature of the liquid oxygen and the high temperature of the flame (about 3500°C) and the very powerful high speed pumps required, the engine itself is very complicated.

The Saturn V rocket used in the first moon landing. It had three parts; each part was dropped when its fuel was used to reduce weight. The first part is shown falling away. Saturn V was 150m tall, weighed 2700 tonnes, and gave a thrust of over 30 000 kN.

Unusual engines

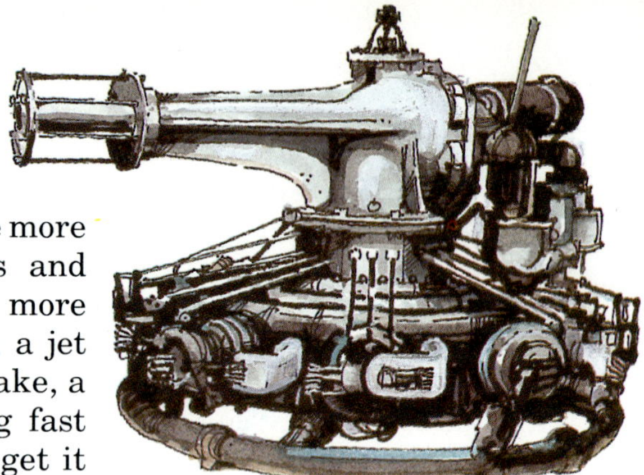

There is unfortunately no room in this book to describe more than a few of the fascinating ideas of engineers and inventors that have led to the production of the more unusual engines. There is the *ram jet*, for instance; a jet engine with no turbine or compressor, just an air intake, a combustion chamber and a jet nozzle. If it is going fast enough the air compresses itself. The problem is to get it started! A similar engine, with valves to make it work in pulses, was used in the German 360 mph V1 flying bomb of World War II.

The strange Canton Unné aero engine.

When the Americans came into the war they needed a lot of tank engines quickly. Chrysler Company had the brilliant idea of sticking five motor car engines together to make one tank engine.

A less successful idea was the 1912 Canton Unné radial aero engine shown, with its horizontal cylinders and vertical crankshaft; it is difficult to see what the designer had in mind, and the big advantage of a radial engine was lost as it had to be water-cooled.

Ettore Bugatti, designer of some of the most elegant and successful racing cars made, also designed a luxury car in 1927 so enormous and expensive that only 7 were made; however, the huge 12.8 litre petrol engines were used in rail cars, with great success, breaking several speed records.

The Wankel engine has no pistons or cylinders, but a specially shaped chamber and rotor produce the same effect. Petrol consumption is rather high; but operation as a steam or compressed air engine is also being tried.

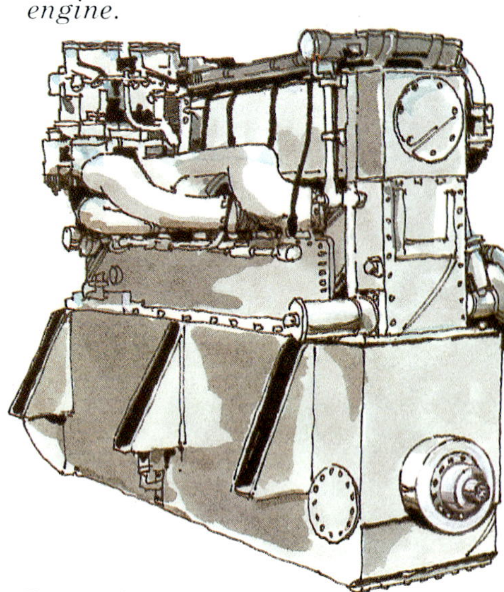

Bugatti's enormous engine for his Royale motor car. It was nearly 1 m high and had eight cylinders in line.

The Wankel engine. *For simplicity, the diagrams show only what happens to one charge; there are three charges in the engine at any one time, all at different stages.*

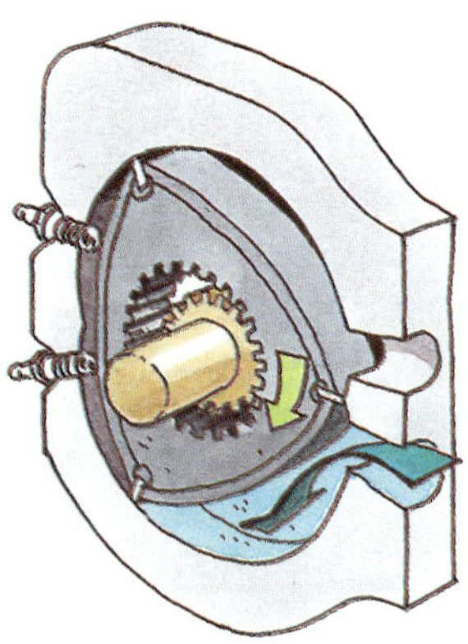

(1) Air and fuel are drawn in and compressed.

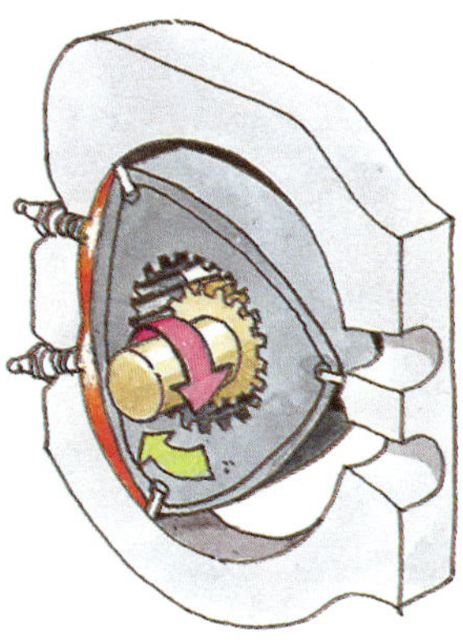

(2) The sparking plugs fire the mixture.

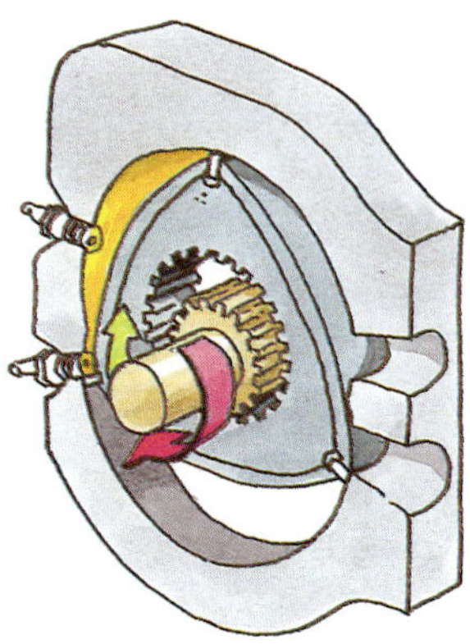

(3) The power stroke.

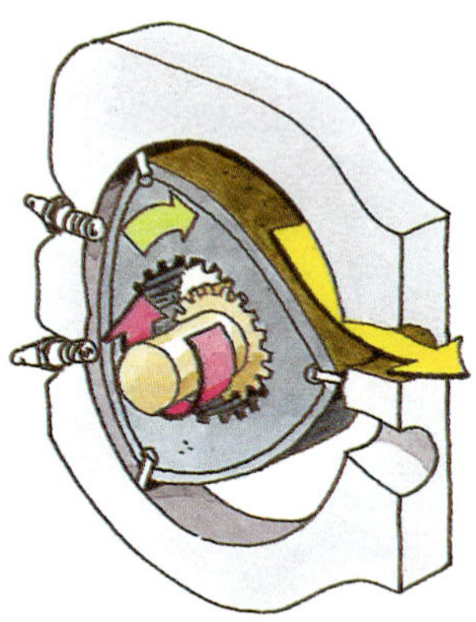

(4) Exhaust.

The future

The world depends on engines – without them there would be no cars, lorries, aeroplanes, trains, or tractors. Unfortunately, engines need oil, and there is only a limited amount in the world. It seems likely that oil will be much scarcer within 40 or 50 years, but coal supplies could last much longer, say 300 years. There are many ways in which this serious problem for the future can be tackled.

Already sugar cane is being grown in Brazil to produce alcohol which can replace petrol. Oil can be obtained from coal, but this is still a very expensive process. Hydrogen gas obtained from water is a possible fuel for the future.

Engines are being made more efficient by using micro-computers to control the ignition, the cooling and the fuel flow. Cars are being made lighter and more streamlined to reduce fuel consumption. The exhaust gas, instead of being wasted, can be used to drive a turbine. Rigid "sails", made like aeroplane wings, have been tried out on merchant ships, reducing oil consumption by as much as 40%. Locomotives with Stirling engines would be able to burn coal, and be much more efficient than the old steam locomotives. There may be diesel-electric lorries as well as locomotives; diesel engines may be combined with gas turbines or even with steam engines to give greater efficiency.

Engineers and scientists are working hard on these projects today, and they will surely have solved these problems before the natural oil and gas in the ground has run out; after all, they have always solved, eventually, the problems of the past.

This experimental Austin Rover car does not look very different from present designs, but it has much lower air resistance.

BMW racing car engine, with the turbocharger at bottom right.

Shin-Aitoku-Maru
The world's first sail-assisted oil tanker was constructed in Japan in 1980.

Time Chart AD 1600–1700

Far East	Middle East	Europe

1600 East India Company formed

1550 Steam-driven roasting spit

1620 Pilgrim Fathers. Mayflower voyage

1629 Branca's turbine

1640

1641–1645 English Civil War

1654 von Guericke demonstrates atmospheric pressure

1660

1679 Papin's safety valve

1680

1680 Huygens' gunpowder "engine"

1698 Savery's "Miner's Friend"

1700

Time Chart AD 1700–1800

Britain | America

Year	World events	Technology
1700		
	1707 Act of Union between England and Scotland to form Great Britain	
		1712 Newcomen's first engine
1720		
	1724 Missionaries expelled from China	
1740		
	"Age of Enlightenment" in Europe (1694–1778 Voltaire, 1712–78 Rousseau)	
	1757 British victories in India (Plassey)	
1760	1759 and Canada (Quebec)	1760 Rockets used in India
		1765 Watt invents the condenser
	1768 Cook's first voyage to the Pacific	1768 Smeaton's experiments on efficiency
		1769 Cugnot's steam tractor
		1776 First Boulton and Watt engine
1780	1776–83 American War of Independence	
		1783 Watt introduces the horsepower
		1786 Evans experiments with high pressure steam
	1788 Colonization of Australia begun	
	1789 French Revolution	
1800	1799–1815 Napoleonic Wars	1800 Trevithick's first high pressure engine

Time Chart AD 1800–1900

Far East | Middle East | Europe

1800		
		1803 Steam boat "Charlotte Dundas II"
		1806 Steam boat "Clermont"
		1812 Trevithick's traction engine and Cornish boiler
		1813 Congreve's rockets at Battle of Leipzig
	1815 Battle of Waterloo	1814 "The Times" printed by steam
		1816 Stirling's hot air engine
1820		
		1824 Carnot's "On the Motive Power of Fire"
		1825 Stockton and Darlington Railway
		1829 The Stephensons' "Rocket"
	1837–1901 Victorian Era	
1840	Colonization of New Zealand begins	
		1843 Joule's "The Mechanical Value of Heat"
		1843 Brunel's "Great Britain"
		1844 Fairbairn's Lancashire boiler
	1848 Revolutions in Europe	1845 McNaught's compound engine
	1848 Californian Gold Rush	
		1851 Lord Kelvin's "On the Dynamical Theory of Heat"
		1859 Rankine's "Manual of the Steam Engine"
1860	1861–65 American Civil War	1860 Lenoir's gas engine
		1861 Babcock and Wilcox water tube boiler
		1862 Beau de Rochas patents four-stroke cycle
	1870 Franco-Prussian War	
		1874 Brayton's "Ready Motor"
	1876 Queen Victoria created Empress of India	1876 Otto's four-stroke gas engine
		1878 Dugald Clerk's two-stroke gas engine
1880		1881 Dowson's gas producer
		1884 Parson's first turbine
		1885 Benz tricycle
		1886 Daimler car
		1890 Akroyd Stuart's oil engine
		1891 de Laval's impulse turbine
		1893 Maybach's carburettor
		1897 "Turbinia" at Naval Review
1900	1899–1902 South African or Boer War	1897 Diesel's oil engine

Time Chart AD 1900–1980

Britain | America

Date	World events	Developments
1900		1903 Tsiolkovsky's book on space travel
		1903 Wright brothers' aero engine and first powered flight
		1904 First diesel-powered ships in Russia 1907 "Gnome" rotary aero engine
	1914–1918 World War I	
	1917 Russian Revolution	
1920		1919 Goddard's book on rockets
		1923 Ricardo's "High Speed Internal Combustion Engine" 1924 Bristol turbocharged aero engine
	1929 Wall Street Crash	1929 Diesel locomotives in Denmark
		1930 Whittle's patent for a gas turbine
		1939 Jet-propelled Heinkel He 178 in Germany
1940	1939-45 World War II	1941 Jet-propelled Gloster E28/39 in England
		1944 V2 rockets land on London
	1949 Chinese Revolution	
		1956 Wankel engine 1957 Russia's "Sputnik", first space flight
1960		1960 British Rail's last steam engine "Evening Star"
		1961 Yuri Gagarin, first man in space
	1969 First landing on the Moon	1969 Neil Armstrong, first man on the Moon
		1975 Fuel produced from sugar cane in Brazil
1980		1980 First sail-assisted oil tanker in Japan

Glossary

Atmospheric engine – one which operates by atmospheric pressure on the outside of the piston, with a vacuum in the cylinder.
Atmospheric pressure – the pressure exerted at any point by the weight of the air above it.
Bar – a unit of pressure measurement, roughly equal to the average atmospheric pressure.
Beam engine – one with a heavy overhead beam pivoted at its centre. The piston operated one end of the beam, and the pump rod or connecting rod was worked by the other.
Beer-engine – a hand-operated machine for drawing beer up from a cellar.
Camshaft – the shaft carrying the cams which operate the valves in an internal-combustion engine.
Carburettor – a device for controlling the mixing of fuel with air as they enter a petrol engine.
Combustion chamber – the part of an internal-combustion engine in which the fuel is burned.
Compounding – designing an engine as a compound engine.
Condense – to change steam into water by cooling it.
Condenser – the part of a steam engine in which the steam is condensed into water so that it can be used again.
Cornish boiler – a type developed in Cornwall with one large fire tube.
Cornish engine – the name given to the Watt type of engine after its development by Trevithick in Cornwall.
Crankshaft – the rotating part of the mechanism for changing the to-and-fro motion of the piston into rotation.
Cycle – in mechanical engineering, the series of operations that take place in the cylinder of an engine.
Cylinder – the part of an engine in which the piston moves; the part of a steam turbine in which one of the rotors is housed.
Cylinder head – the part that closes the end of the cylinder of an internal-combustion engine, forming part of the combustion chamber.
Diesel engine – an oil engine in which the air is compressed so much that it is hot enough to set fire to the oil when it enters the cylinder.
Displacer – the piston and cylinder which pump the air from the cold end to the hot end of a hot air engine, and back again.
Double acting – an engine in which steam acts on both sides of the piston in turn.
Efficiency – the useful output of an engine, divided by the energy in the fuel supplied to it.
External combustion engine – one in which the fuel is burned outside the cylinder, as in a steam engine or a hot-air engine.
Fire tube – a tube in a boiler carrying the hot gases from the furnace through the water.
Fitter – a highly skilled mechanic who can carry out many jobs in the building or repair of engines and mechanisms.
Flywheel – a heavy wheel fitted to an engine so that it doesn't slow down or speed up too much as the force on the piston varies during the cycle.
Four-stroke engine – an internal-combustion engine in which two revolutions of the crankshaft are required for one power stroke in a cylinder.
Gas turbine – an internal-combustion turbine, burning a fuel similar to paraffin.
Governor – a mechanism for controlling the speed of an engine automatically by increasing or decreasing the supply of steam or fuel as the load on the engine increases or decreases.
"Grasshopper" engine – a beam engine in which the beam is not pivoted at the centre but controlled by a swinging rod at one end.
Horsepower – a measure of power, equal to 0.746 kilowatts.
Hot tube or hot bulb ignition – a method of igniting the fuel in an oil engine by conduction of heat along a tube from an external flame.
Ignition system – the system for setting fire to the fuel in an internal-combustion engine – apart from a diesel engine.
Impulse turbine – one in which the steam loses speed rather than changing direction as it hits the turbine blades, as in Branca's machine.
Injection – forcing the fuel into the cylinder after the air has been compressed, instead of allowing it to be sucked in with the air.
Internal-combustion engine – one in which the fuel is burned inside the engine, as in a petrol, oil or gas engine, or gas turbine.
Jacket – a container surrounding the cylinder of an engine, with water or steam for cooling or heating.

Jet engine – one in which the exhaust gases flow freely out of a nozzle at the rear of the engine. Gas pressure on the front of the inside of the engine produces a forward propelling force.
Knot – the measure of speed used in ships and aircraft. One knot is one nautical mile per hour, equal to 1.15 land miles per hour, or 1.85 metres per second.
Lancashire boiler – a type developed by Sir William Fairbairn in Lancashire, having two fire tubes instead of one.
Locomotive – an engine which can propel itself along on land; usually applied only to railway engines nowadays.
Magneto – a device for producing a spark without the use of a battery. It has a rotating coil driven by the engine between the poles of a horseshoe magnet.
Opposed piston engine – one in which the cylinders do not have closed ends. There are two pistons working in opposite directions. The combustion chamber is the space between the two pistons where they nearly meet in the middle of the cylinder.
Piston – the part of an engine which slides to and fro in the cylinder, and on which the expanding gases exert their force.
Radial engine – one with the cylinders arranged around a central crankshaft, rather like the petals of many flowers. Many air-cooled aero engines were made like this.
Ram jet – a jet engine with no turbine or compressor, relying on its own speed through the air to compress the air entering it.
Reaction turbine – one in which the steam changes direction rather than speed to make the rotor turn, as in Hero's turbine.
Rpm – the usual measure of the speed at which an engine is working.
Rotative engine – a beam engine which drove a shaft by means of a connecting rod and a crank or gear wheels.
Rotor – the rotating part of a turbine.
Safety valve – a valve which opened if the steam pressure in a boiler became too high to be safe.
Side lever engine – a beam engine with a beam brought down to the side of the cylinder to reduce the height. There were usually two levers, one on each side of the cylinder.
Smoke box – the space at the front of the boiler of a steam locomotive into which the furnace gases and the exhaust steam from the cylinders were led before leaving through the funnel.
Stroke – the distance moved by the piston in the cylinder from one end of its travel to the other, while the crankshaft turns through one half of a revolution.
Supercharger – a pump for forcing the air into the cylinder of an engine, as opposed to letting it flow in by atmospheric pressure.
Superheated steam – steam that has been heated above the boiling point of water.
Supersonic – a speed greater than the speed of sound. The speed of sound in air at sea level is 340 m per second (760 miles per hour).
Thermodynamics – the science of heat and its use in engines.
Triple expansion engine – a compound engine in which the steam passes through three cylinders in turn – high pressure, intermediate pressure, and low pressure.
Turbine – an engine in which the flow of gas or steam causes a rotor to turn by acting on many blades, rather like the wings of an aeroplane, arranged around the outside of a disc. There are also water- and air-driven turbines.
Two-stroke engine – an internal-combustion engine in which each cylinder gives a power stroke in every revolution of the crankshaft.
Vacuum – strictly speaking, a space in which there is nothing at all. Usually though it means a space in which there is a little gas or air at a pressure well below atmospheric.
Valve – a device for controlling the flow of gas, liquid or electric current.
Watt – the measure of power. 1000 watts is 1 kilowatt; an electric fire has a power of 1 to 3 kilowatts.
Work – in engineering, work is said to be done by a moving force; for example, in lifting a weight

Index

Aero engines 4, 5, 28, 31, 35–37, 40
Aircraft 4, 5, 25, 28, 35–37, 40, 41
Atmospheric pressure 6–9, 11, 26, 27

Babcock and Wilcox boiler 19
Beam engines 8–15, 18, 21
Benz, Berthe 28
Benz, Karl 28, 29, 30
Boilers 6, 7, 13, 19, 21, 23, 26
Bosch, Robert 31
Boulton, Matthew 11, 15
Branca, Giovanni 6, 22
Braun, Wernher von 39
Brayton, George 32
Brunel, Isambard 21
Bugatti, Ettore 40

Camshafts 29
Carburettors 30, 31
Carnot, Sadi 20, 33
Cayley, Sir George 25
Clerk, Sir Dugald 33
Compound engines 18, 21
Condensers 11, 13, 18, 21, 23
Congreve, Sir William 38
Cugnot, Nicholas 16

Daimler, Gottlieb 27–30
Darby, Abraham 9
Diesel engines 5, 21, 32–35, 41
Diesel, Rudolf 32, 33
Displacers 24, 25
Dowson, Joseph 27

Ericsson, John 25
Evans, Oliver 13–15

Ford, Henry 31
Four-stroke engines 26, 27, 29, 33, 34, 36
Fulton, Robert 14

Gagarin, Yuri 39
Gas engines 5, 26–28
Gas turbines 5, 36, 37, 41
Gloster E28/39 36, 37
Goddard, Robert 39
"Grasshopper" engines 14, 15
Guericke, Otto von 6

Heat pump 20
Hero of Alexandria 6, 22
Hot air engines 5, 20, 24, 25, 41

Ignition systems 30–33, 41
Impulse turbines 22
Industrial Revolution 4, 9
Internal-combustion engines 5, 24, 26–35, 41

Joule, James 20

Kelvin, Lord 20

Langen, Eugen 26, 27
Laval, Gustav de 22
Lenoir, Etienne 26, 27, 30, 32
Locomotives 5, 13, 16, 17, 19–21, 25, 33, 34, 36, 40, 41

McNaught, William 18
Magnetos 31

Maudslay, Henry 14, 21
Maybach, Wilhelm 27, 30
Mercedes-Benz 29

Newcomen, Thomas 8–10, 12, 19, 21

Ohain, Hans von 36
Otto, Nikolaus 26, 27, 29, 30

Papin, Denis 8, 10, 14, 26
Parsons, Sir Charles 22, 23
Petrol engines 5, 28–35, 40, 41
Philips 25
Planet locomotive 17
Portable engines 15
Priestman family 32

Railways 5, 16, 17
Ram jets 40
Rankine, William 20
Reaction turbines 22, 23
Refrigerators 20, 32
Ricardo, Sir Harry 34
Rochas, Beau de 26, 27
Rocket locomotive 17, 19, 25
Rockets 22, 36, 38, 39
Rolls Royce 4, 35, 37, 38

Safety valve 8
Savery, Thomas 6, 7, 8, 9, 14
Ships 14, 15, 21–23, 25, 26, 34, 36, 37, 41
Side lever engines 14, 15
Smeaton, John 9, 10
Steam engines 4, 5, 8–22, 24, 26, 28, 41
Steam turbines 5, 6, 22, 23, 36, 37
Stephenson, George and Robert 16, 17, 19, 25
Stirling, Patrick 17
Stirling, Rev. Robert 24, 25, 41
Stuart, Akroyd 32, 33
Superchargers 35, 41
Superheating 21
Symington, William 14

Thermodynamics 20
Times, The 16
Trevithick, Richard 13–17, 19, 22
Tsiolkovsky, Konstantin 39
Turbinia 22, 23
Two-stroke engines 33, 36

Valve gear 7, 9, 12, 18, 26, 28, 29, 33, 35

Wankel engine 40
Watt, James 10–15, 19, 21
Whittle, Sir Frank 36, 37
Wilkinson, John 12